Health Informatics on FHIR: How HL7's New API
is Transforming Healthcare

Mark L. Braunstein

Health Informatics on FHIR: How HL7's New API is Transforming Healthcare

 Springer

Mark L. Braunstein
School of Interactive Computing
Georgia Institute of Technology
Atlanta, GA, USA

ISBN 978-3-030-06655-0   ISBN 978-3-319-93414-3   (eBook)
https://doi.org/10.1007/978-3-319-93414-3

This Springer imprint is published by the registered company Springer International Publishing AG part
of Springer Nature.
The registered company address is: Gewerbestrasse 11, 6330 Cham, Switzerland

"We will be laser-focused on increasing interoperability and giving patients access to their data. Last year, CMS finalized requirements around EHR certification. This ensures that patients will be able to share data via APIs"

– CMS Administrator, Seema Verma,
HIMSS 2018

I write this book as I near the end of my 7th decade. This has been a time for reflection, in part because we have recently lost so many of the giants of this field. Some of these people were of great importance to me. Even as I was doing the research for the book, we lost Larry Weed. His book Medical Records, Medical Education and Patient Care[1] opened my eyes to the possibilities of computerized patient records and inspired me to work on an early ambulatory EMR based on his problem-oriented charting principles. Three years earlier, we lost Morris Collen who was instrumental in introducing me and my first company to Kaiser.

While the contributions of the pioneers I profile in this book have been well chronicled elsewhere, those books and articles are often written narrowly for our health informatics community, either because of their technical nature or the venues in which they are published.

---

[1] The Press of Case Western Reserve University, Year Book Medical Publishers, 1969.

As in my prior books, it has been my goal to make health informatics more widely accessible to medical students, physicians and other health professionals, patients and nontechnical readers as well as software developers wanting to learn about this rapidly evolving and growing field. I know from the many e-mails and chance meetings over the years at conferences that my books have often been of service to my intended audience. I hope that this new edition will be equally accessible and will also introduce readers to some of the people who built this field and to their pioneering efforts.

However, the one person I wish to dedicate this book to is thankfully still very much with us! I suppose that most of us dream about changing the world but, of course, very few actually accomplish that. Grahame Grieve is the founder of HL7 FHIR and the FHIR Product Director at HL7 International. I believe he is having a virtually impossible to measure positive impact on health informatics. As a result he is doing as much as anyone to create the safer and more clinically productive and cost-effective global healthcare system that many of us have devoted our careers to. Reading his 2011 post introducing "Resources for Health" (that became HL7's new Fast Healthcare Interoperability Resources (FHIR) standard) it was clear to Grahame from the outset where things needed to go.[2] I will quote from that seminal post later on but let me briefly describe Grahame whom I consider a friend despite the vast distance between Melbourne and Atlanta.

---

[2] http://www.healthintersections.com.au/?p=502

I had admired Grahame's work from a distance but it was only when my close public health colleague, Paula Braun, and I arranged for him to give lectures on FHIR at Georgia Tech and at the Centers for Disease Prevention and Control (CDC) that I realized just how brilliant unassuming Grahame actually is.

As is often the case with innovation, Grahame came along at precisely the right time with exactly the right ideas. What is not so often the case is that the idea generator also has the people and organizational skills to guide a diverse global group of volunteers toward the desired result.

I am sure that I am not the only longtime member of this community who never expected to live to see the dramatic transformation that is taking place in large part because of FHIR.

Grahame, we are lucky to have you in the health informatics community, and thanks for all you are doing to move us ahead. I am thrilled to be able to dedicate this book to you and please do not be too angry with me for doing it.

Foreword

I often kid Ed Hammond that I am grateful he is still around because, if he were not, I might well be the oldest person in our field! Well into his 80s Ed maintains a schedule that would tax a much younger person. We first met in 1973 when I was supported by the Duke Foundation to study the British healthcare system over a wonderful summer in London. Ed was one of the instructors. His training up through his Ph.D. was at Duke University. He never left and today, among other leadership positions there and elsewhere, he is Director of the Duke Center for Health Informatics. His career is well described by the presentation speech when he received the highest honor in our field, the 2003 Morris F. Collen Award of Excellence from the American College of Medical Informatics.[1] Given the focus of this book, it is particularly important to point out that his interest and involvement in health standards literally goes back to "day 1" when, in 1983, he began work with another health informatics pioneer, Clem McDonald, and others to create messaging standards for exchange of data among systems, the predecessor to today's HL7 messaging standard.

Comments

Dr. Mark Braunstein asked me to share some of my insights on this book. First, I like the book. Mark and I have shared experiences over much of the period of the information age and particularly its use and impact on health care. The book is interesting in its glimpses in the early history and significant focus on what is happening today. Reading Mark's snippets of history, I began to realize that history is defined by writers and what they think is important. Even the most extensive documentation of history leaves out more of history than what is included. I would include different events of history, such as the early work of Lockheed and El Camino Hospital as the beginnings of Hospital Information Systems (HIS) and of

---

[1] https://www.ncbi.nlm.nih.gov/pmc/articles/PMC400521/

Octo Barnett at Massachusetts General Hospital and the development of MUMPS and COSTAR as one of the early EHR systems. But then, I am defining history as I see it. As I share these insights, I recognize the importance in history that influenced the development and implementation of standards over the history of time.

For both of us history is the transition of technology from a point in which it was the limiting factor to where it now becomes the innovator and enables the vision. We both worked in the world of minicomputers with limited main memory (4 K bytes of main memory), slow computing speeds, slow output speeds (110 baud), cumbersome output devices (teletypes with cloth ribbons and continuous rolls of paper and character-based terminals), to very limited connectivity. The amazing exponential technological progress in all of these areas is important in looking at what the future may cover. Dr. Braunstein does an excellent job of marking this progress from the past to looking at the future.

Health Level Seven came into being as a result of a need of Don Simborg and his company to develop a HIS from a cadre of niche vendors. This requirement was met by creating a standard with which data could flow among systems in a standard fashion. Although the basic concepts were defined from previous work done by Simborg, the work by Clem McDonald and others with one of the first laboratory standards influenced the actual format and syntax. Strong arguments among the developers were whether we used tags or delimiters. The latter won, because of the cost of transmitting each character was high. The "suggested" delimiters were taken from The Medical Record, developed at Duke University.

I think the discussion of the early standards developing organizations (SDOs) is important. Each of the seven or so main SDOs were created for different purposes and not intended to be competitive. However, as the scope of each SDO expanded, scope spread and overlaps occurred. HL7 v2 was not built around X12 syntax. In fact, one of the early battles was between X12 calling itself ANSI, and HL7 was not initially accredited by ANSI. There was a struggle to get HL7 to become ANSI qualified, but importantly that happened and has been important to HL7 over time.

One of the great achievements of this book is explaining a complex, technical topic in a manner that is more easily understandable by lay people. This book talks about how and when to use a standard, rather than the technology of standards or who makes the standard. The breadth and innovative use of these standards will result from clinicians and researchers understanding what can be accomplished with these standards more than the technical community trying to guess at appropriate applications.

Dr. Braunstein covers the broad range of standards. His footnotes and URLs provide key links for those wishing to learn more about the topics. He identifies key persons in the standards community. He provides many real world examples.

I am impressed with the explorative work of Dr. Braunstein in the use of FHIR. FHIR has been given more hype than any other standard and probably more than any other functionality in health information technology. FHIR will succeed as a result of how it is implemented as well as the functionality of the standard. The flexibility of the standard with extensions to resources may defeat the simplicity of FHIR if used improperly. There is already some push-backs to FHIR and concerns from the lack of maturity in the standard. Dr. Braunstein, with his extensive research, has identified a number of successful implementations of FHIR within major institutions and a number of start-ups. People also need to realize that a static standard is an out-of-date standard and that FHIR and other standards will continue over time to meet new requirements.

In conclusion, I like this book. It meets the need of an intended audience, and I plan to use the book as a reference for a course I am teaching. Well done, Mark.

W. Ed Hammond, PhD
Director, Duke Center for Health Informatics, Duke Clinical & Translational Science Institute
Director, Applied Informatics, Duke Health Technology Solutions
Director, Academic Affairs, MMCi Program, Duke School of Medicine
Professor, Community and Family Medicine
Research Professor, School of Nursing
Professor Emeritus, Biomedical Engineering
Duke University
Chair Emeritus, Health Level Seven International
Durham, North Carolina

Foreword

Within these pages, we start with a most gentle introduction to twentieth century healthcare informatics. As he paints this complex landscape, the reader is at once struck that Dr. Braunstein is in love with his subject. The timeline that he describes is torturous, and the subjects do not ever seem to sit still for the artist. Instead, the reader is led through a challenging and often byzantine world that is hesitantly called information technology. Although we may conclude that we have reached this time of remarkable innovation simply by chance, the author provides an impeccable story, even if not all of the most critical inflection points were as effortless as he has described them.

Dr. Braunstein has a vision that has evolved over four decades, sometimes with adolescent certainty, others with the wary eye of a school teacher. But, teacher he is, and he is as proud of his students as he is of his mentors. As we enter his narrative, the modern realm of computer science seems to gently emerge, rather than leap upon us. It is clear from the very start that he is describing a critical change in the sharing of healthcare information, even while the moth is being transformed into a butterfly.

The drive for better healthcare certainly requires much more than technology. As you read about the impact of information science on our patients, we must not forget that medical errors are the third leading cause of death in the United States. For many of us, there is the hope, so elegantly described within these pages, that we have reached a point of no return. The cost of healthcare is approaching twenty percent of the gross domestic product, but improvement in healthcare quality for very many does not reflect that staggering financial investment we have made. Millions of us go from day to day without any semblance of prevention and limited access to care when we become acutely ill. Our systems continue to focus on episodes of care, rather than management of chronic illnesses. Moreover, our healthcare system is built upon payment for services rather than rewarding positive outcomes.

In the author's eye, the sciences of information technology and medical informatics will serve as a transformational force to reframe this discussion. When *Health Level 7* embarked on a distinctly different approach to information sharing, the informatics community barely took notice. This remarkable innovation bears the acronym of *FHIR*. The adoption of the Fast Healthcare Interoperability Resource platform has emerged over 8 years as the change agent that Dr. Braunstein so passionately details. This process required the contribution of a true international community of technical developers and clinical implementers. Every step has left its mark.

Very early decisions by the FHIR leadership were trustingly adopted. Partnerships were formed, without which the gradual but steadily growing adoption of FHIR could not have been achieved. The critical value of these collaborations is prominently highlighted within these pages. The essential support of the Argonaut Project and SMART Health IT from Boston Children's Hospital provided resources and leadership at a crucial time in the infancy of FHIR.

As this volume is about to go to press, the global adoption of FHIR is now accelerating. Government agencies in the USA have recognized the essential role that FHIR will begin to play as we drive toward true personalized medicine and dramatically enhanced care delivery, wellness, and patient engagement. True patient-centric care is no longer like the unicorn that everyone talks about but no one has actually seen. Within implementation communities around the world the integration of genomic data at the point of care is being demonstrated and enhanced clinical decision support is making everyone a better clinician. At the Veterans' Administration, the Lighthouse Gateway Project has made an ambitious promise to bring timely information and better care to those who have sacrificed for all of us.

The private sector continues to support FHIR innovation with both FHIR application programming interfaces (APIs) and with FHIR app stores. Many see a major transformation in which the traditional electronic health record system enables the integration of agile and innovative applications that produce the outcomes that the seamless exchange of information has promised. Most recently, Apple Corporation introduced a Health Record application, in which FHIR is integrated into the operating system, and which provides patients with simplified access to their longitudinal health records from different providers and different health systems. To date, nearly four dozen major health systems across the USA have announced their participation in this initiative.

Next year, Dr. Braunstein will almost certainly have important new evidence to add to his narrative. Even now, the payer community, laboratory researchers, and population health scientists find that FHIR provides them with a more comprehensive and more agile means of achieving their goals. Future chapters promise to be both long and revealing ones. For now, the reader is best advised to buckle up and begin what is certain to be a most exciting ride.

Charles Jaffe, MD, PhD
CEO
Health Level 7 International
Ann Arbor, Michigan

Preface

This is the third book in a series I have written in an effort to make contemporary health informatics both accessible and interesting to nontechnical readers, including healthcare providers and their patients and to computer scientists interested in an introduction to the field.

The title of the first book, *Health Informatics in the Cloud*, suggested where I thought the field was headed – toward the use of modern technologies to make healthcare data more accessible, sharable, analyzable at scale, and ultimately more useful for patient care. In 2012, when that book was published, the industry was just beginning to implement the Obama administration's Health Information Technology and Economic and Clinical Health Act of 2009 (HITECH). The program was designed to foster adoption of electronic health records (EHRs). It should be obvious from the title of *this* book that I place great importance on HL7's new Fast Healthcare Interoperability Resources (FHIR) standard. Its development was barely underway when I wrote the first book, and I was unaware of FHIR at the time.

By 2014 when I wrote the second book, *Practitioner's Guide to Health Informatics*, three things were clear. First, the USA would achieve widespread EHR adoption. Predictably, the commercial EHRs that were being adopted had many usability issues and did not readily share information or make it available for novel uses. The term for the missing ability to share data is "interoperability" and today it is at the center of many health informatics technical and policy discussions. Finally, FHIR seemed to offer the real promise of providing interoperability (or, at least, its first two layers – more on that later on in the book) and also providing a long sought "universal health app" platform that could bring innovative uses of EHR data to practicing providers and their patients independently of the underlying EHR.

Today, as I prepare this third book, because of widespread FHIR adoption and the increasing ecosystem of SMART on FHIR apps, we are now in the early years of a nearly complete transformation of once proprietary, closed EHRs into what could become platforms for innovation. This is not a *fait accomplie,* there are important remaining challenges, but significant progress has already been made and FHIR adoption is expanding at what a short time ago would have been an unthinkable rate.

FHIR has now been embraced by key parts of the US federal government including CMS, the agency that runs Medicare which is the largest US healthcare payer and by the Veteran's Administration which runs the nation's largest health system. As we will see, commercial health insurance companies are exploring many opportunities to use FHIR to facilitate their business processes particularly when they require electronic patient record data. Google has launched an ambitious health analytics effort, which makes its data store of health data available using FHIR. Apple is leveraging FHIR to provide a means for iPhone users to aggregate all their medical records and is positioning the phone as a FHIR app platform. Just as this book was going into production Microsoft Healthcare announced that it has hired Josh Mandel as its Chief Architect where his role will be "to lay the groundwork for an open cloud architecture to unlock the value of healthcare for the entire health ecosystem." Given Josh's leadership in its development, it is safe bet to assume his work will leverage FHIR. FHIR is now supported by the largest commercial EHR systems and is increasingly being used in the private insurance sector to connect to providers as they deliver care in order to increase their efficiency and effectiveness. Of perhaps even more importance, FHIR is rapidly becoming the platform for innovation in what has for a long time been a relatively slow-moving field.

There is more to come. FHIR based web services providing expert advice or guidance on a wide range of topics have already started to appear and their number will likely grow exponentially. It is also likely that, as the standard further evolves and grows, it will become important to the representation and optimization of clinical workflow and process. Finally, FHIR may expand from a facile means of accessing data for the care of one patient at a time to a platform for public and population health as well as for the analysis of large scale health "big data."

Despite what I feel will be its ultimate importance, FHIR is still not widely understood or appreciated beyond the health standards and informatics communities. It is my hope that this book will make a broader community of readers aware of this important new technology and how it might impact or be used by them. While I am sure there are important ramifications and applications of FHIR that I am not yet aware of, it has been my goal to cover its current and likely future use as broadly and comprehensively as possible. I apologize to my readers and those whose efforts I should have discussed for any worthy efforts that I failed to include.

While I usually explain technical terms in place, please refer to the extensive glossary should I use or not explain in context a term or concept that is unfamiliar to you. The book is extensively referenced and cites many good sources of additional information. I refer readers wanting an even more detailed discussion of FHIR and interoperability standards in particular to *Principles of Health Interoperability* written by Tim Benson who led the first European project team on open standards for health interoperability and Grahame Grieve.[1]

Atlanta, GA, USA   Mark L. Braunstein

---

[1] http://www.springer.com/gb/book/9783319303680

Acknowledgments

I could not have written this book without the help of a number of friends, colleagues, and interested employees of the companies I profile in it.

My introductory health informatics graduate seminar in the College of Computing here at Georgia Tech occasionally attracts an MD/Ph.D. student, but it was a first when Dr. Dennis Steed, a retired endocrinologist, showed up on the first day of class one semester. Dennis has a real passion for informatics so we have stayed in touch. When I asked him to review the book I had no idea that he would be such a rich source of ideas as well as a great editor. During the course of our back and forth about the book I realized just how well he knows the field but even that did not prepare me to learn that his practice had tried to use INTERNIST-II "back in the days"! They used Quick Medical Reference (QMR), a tool for making diagnoses based on INTERNIST that was developed by Randolph A. Miller, MD, at the University of Pittsburgh in 1980.[1] Dr. Miller is now a Professor of Biomedical Informatics at Vanderbilt University.[2] More about that INTERNIST as we proceed but, for now, thank you Dennis.

Paula Braun is the first Entrepreneur in Residence at the CDC here in Atlanta. She works in that role to "tackle a critical problem with skilled innovators looking to make a meaningful impact and who can solve that problem."[3] Early on, she took my graduate seminar's Udacity MOOC and e-mailed me about it. I suggested we meet and we have been colleagues and collaborators ever since. With Paula's help my graduate students have done around three dozen FHIR app projects that have exposed various experts at CDC to the potential uses of the technology in public health. She provided me with numerous great suggestions and spotted quite a number of organizational as well as grammatical and spelling errors in the book. Thank you Paula.

---

[1] http://www.openclinical.org/aisp_qmr.html

[2] https://www.vumc.org/dbmi/person/randolph-miller-md

[3] https://www.hhs.gov/idealab/eir-program/

Dr. Eric Dahl, Associate Dean for Strategic Initiatives at the University of Georgia's College of Public Health, reviewed Part I of the book and made many valuable suggestions. Lucie Ide, MD, PhD, the founder of a FHIR based company profiled later on, read the entire book and both found mistakes and made many useful comments.

Thanks to these people who reviewed sections of the book for accuracy: Grahame Grieve, Ken Mandl, David Hay, Josh Mandel, James Cimino, Peter Celano, Alistair Erskine, Ted Shortliffe, Gregory Cooper, and my Georgia Tech colleagues Jimeng Sun and Jon Duke.

In preparing the book, I identified several academic and commercial organizations that I felt were good exemplars of innovative health informatics, often involving current or potential future use of FHIR. In each case, one or more people were kind enough to work with me to craft an accurate discussion of what their organizations were doing.

Apervita: Blackford Middleton, MD, Chief Informatics and Innovation Officer
CareEvolution: Michele Mottini, Software Developer
CommonWell Health Alliance: Jitin Asnaani, Executive Director
Continua: Michael J. Kirwan, Vice President, Continua, Personal Connected Health Alliance
Da Vinci Project: Viet Nguyen, MD, Clinical Informatics Consultant, and Jocelyn Keegan, Senior Consultant, Point-of-Care Partners
Duke University: Sophia Smith, PhD, MSW, Associate Professor, School of Nursing
Human API: Scott Cressman, Head of Growth and Stephanie Wilson, Director of Operations and Partnerships
Humana: Patrick Murta, Principal Solution Architect
Juxly: Howard W Follis, MD, CEO and President
MEDITECH: Lawrence O'Toole, Associate Vice President, and Joe Wall, Manager Interoperability Strategy, Certification and Health IT Policy
PatientsLikeMe: Margot Carlson Delogne, VP, Communications
Philips Wellcentive: Mason Beard, Chief Solutions Leader and Kirk Elder, Chief Technology Officer
Rimidi: Lucie Ide, MD, PhD, Founder and Chief Health Innovator
Surescripts: Mark Gingrich, CIO and Keith Willard, Chief Architect
TIBCO Software: Matt Creatore, Strategic Account Executive, and Beth Spars, Account Executive,
Utah Health Information Network: Teresa Rivera, President and CEO
Validic: Ashley Rae Needham, Director, Corporate Initiatives
vRad Radiology: Shannon Web, President and COO, Benjamin W. Strong, MD, Chief Medical Officer
Zynx Health: Aslan Brooke, Director, Technology, Security Officer

About This Book

As its title telescopes, this book looks at health informatics through the lens of the new and rapidly evolving Health Level Seven® (HL7) Fast Healthcare Interoperability Resources (FHIR®) standard that I feel is literally transforming the field by solving the long standing interoperability problem and making health data far more usable by healthcare providers and their patients.

This book is divided into four parts. Here I provide an overview and how they relate to the overall discussion.

Part I: These first three chapters discuss some of what I consider to be the most interesting aspects of the history of health informatics. This discussion, in part, also telescopes the historical difficulty of moving informatics concepts into actual patient care. This part also discusses the nature, structure, economics, and key problems of the US healthcare delivery system informatics seeks to serve, and some of the innovations in the care delivery and payment models that offer hope for future improvement. The reader will hopefully see the key roles that informatics plays in creating and successfully operating those new models as well as some of the remaining challenges, particularly with respect to the usability of electronic health record systems. This part introduces the vision of a "Learning Healthcare System" in which digital data from actual patient care is aggregated and analyzed to gain new knowledge that is then fed back to the care delivery process from which the data came in what can fairly be viewed as a, hopefully positive, feedback loop. Finally, it explains at a high level the federal programs to spur health informatics adoption begun in the second Bush administration and more fully funded and vigorously pursued by the Obama administration.

Part II: The first part introduces the most visible health informatics tool, the provider-facing electronic health record. This part's three chapters introduce three key additional opportunities for informatics – patient empowerment, health information exchange, and payment for healthcare services. In each of these, we discuss how health informatics is being used along with some of the challenges it faces and how it will likely be used in the future with a particular emphasis on the role HL7 FHIR may play in that domain. This book largely focuses on contemporary informatics, so the chapters vary significantly in length and detail because of the variation in the scope of informatics efforts, particularly with respect to the HL7 FHIR standard, in each domain. For example, the use of application programming interfaces (APIs), and FHIR in particular, to engage patients is a substantial discussion in part because it is a key federal priority, while adoption of FHIR by health insurance companies is a relatively new development.

Part III: The four chapters in this part discuss the standards that are the foundation of most contemporary and likely future health informatics systems and tools. The first two chapters cover the prior and current approaches to representing and sharing health data. The next two describe the FHIR standard and the SMART on FHIR app platform that is built using that standard in order to make health data more accessible and useful to providers and their patients.

Part IV: The final three chapters describe how informatics, and FHIR in particular, are being used in three of the most dynamic, forward facing, and potentially impactful health domains. The last of these chapters focuses on the final analytics phase of the vision of a Learning Health System with which the book begins.

For me one of the most difficult aspects of writing a book is creating the best sequence for the presentation of the information. At times this creates difficult decisions for which there may be no optimal answer. For example, early on in the chapter on Health Information Exchange (HIE) we discuss semantic interoperability, a technical and somewhat difficult concept that would appear to be better left for one of the later chapters on standards. However, no discussion of HIE would be complete without an examination of the Indiana Health Information Exchange (IHIE), which is arguably the most successful and most sophisticated operational HIE in the USA. A core IHIE achievement is the creation of a central data store in which there is a significant degree of semantic interoperability. I feel it would be unfair to readers to introduce IHIE before they have an appreciation for the nature of semantic interoperability and the difficulty of creating it from complex and often inconsistent and error prone medical data. If you do run into things you do not understand, please remember there is an extensive glossary at the end of the book.

While I always endeavor to make information clear and accessible to all readers, some parts of the book are necessarily more technical than others. In no case is that material critical to less technical readers who should be able to skip ahead without getting confused.

To help put the material into perspective I provide ten commercial case studies of innovative informatics. I also provide five detailed discussions of prototype FHIR apps done by students in my Georgia Tech CS6440 graduate seminar under the direction of domain experts. For clarity these sections are shaded to separate them from the main body of the text.

Finally, throughout this book I provide citations to articles and resources to further supplement your appreciation of the field. In virtually all cases the material is freely available on the Internet, so I provide links. Given the ever-changing Internet landscape, I apologize in advance for any links that no longer work so please report these and I will endeavor to post updated links on the book's website.

Atlanta, GA, USA   Mark L. Braunstein
April 2018

Contents

Part I
Perspective

Chapter 1
A Brief History and Overview of Health Informatics

1.1 Introduction

This first chapter provides a brief history of some, but certainly not all, of the key subdomains within the health informatics field and further explains the potential significance of the FHIR standard that will occupy much of the rest of the book. To do this, the chapter begins with a discussion of early electronic records and clinical decision support tools and then shifts gears to introduce the concept of health information exchange. Later, we discuss interoperability challenges that date back decades and the various ways that existing technologies have been used, sometimes with limited success, to simplify and coordinate the sharing of information among providers. The chapter ends with the premise that widespread adoption of modern web technologies (and FHIR in particular) is transforming health informatics. To help illustrate this the chapter ends with a demonstration FHIR app developed by a team of Georgia Tech students did using these emerging technologies to help predict the onset of a life threatening condition in ICU patients.

One of the things I thought about in preparing the book was the rate at which we are losing the early pioneers in health informatics so I have made a conscientious effort to briefly describe their work wherever it fits into the narrative. In many cases this emphasizes how long-standing many of the key health informatics challenges are, and it enriches the discussion. As I will throughout the book, I provide references or links to supplemental material that should provide more detail and expand your appreciation of the field.

1.2 Early Electronic Records and Clinical Decision Support

This book is about health informatics, which encompasses the many applications of computing involved in the *delivery* of healthcare. This is distinct from the related field of bioinformatics which seeks to understand and model the incredibly complex

© Springer International Publishing AG, part of Springer Nature 2018
M. L. Braunstein, *Health Informatics on FHIR: How HL7's New API
is Transforming Healthcare*, https://doi.org/10.1007/978-3-319-93414-3_1

biochemical engine within each of our cells. Historically health informatics has therefore been about applying technologies rather than gaining new knowledge. This is changing because of the increasing sophistication of modern analytic and machine learning[1] technologies coupled with the broader availability of digital health data from providers and their patients. It is now clearly the case that health informatics can be the source of new knowledge about how best to diagnose and treat disease in real world patients. Readers may also run into the term 'biomedical informatics' which I, perhaps loosely, describe as the combination of these two technologies into one academic discipline and/or department.

The field has a long history. Since the 1950s when computers could first store and process 'large (by the standards of the day)' amounts of data, software developers have seen healthcare as a fertile domain. Electronic medical record systems (EMRs) date back at least to the early 1960s when Akron Children's Hospital and IBM created HIS, which is claimed to have been the first computer-based clinical hospital information system (Fig. 1.1). I provide a link to an IBM video from the period explaining the system.[2] I believe you will find it quite entertaining.

Ambulatory electronic medical record systems for use outside of hospitals began in the late 1960s. Starting in the 1970s, as you can see in the photo of me back then, I oversaw the development and did some of the programming of one of these early EMRs at the Medical University of South Carolina (MUSC) in Charleston (Fig. 1.2).[3]

Clinical decision support systems were another important early application of informatics to patient care. These systems seek to guide physicians based on scientific evidence as they make diagnostic and therapeutic clinical decisions. They date at least to 1972 when Stanford health informatics pioneer, Dr. Ted Shortliffe, began work on MYCIN, a program to assist physicians to make optimal antibiotic selection to treat infections. We will discuss MYCIN in some detail later on in the book but, a number of years later, Dr. Shortliffe posted a pair of YouTube videos to demonstrate and memorialize the program since, as he said, the computers it could run on were disappearing.[4]

---

[1] Machine Learning is the branch of artificial intelligence focused on the development of algorithms to do things such as classifying or clustering items of interest. In supervised learning input and output data are labeled and an algorithm learns the mapping function from the input data to the output labels by comparing its output to the correct output. In unsupervised learning there is only unlabeled input data and so the model is left to discover and present interesting structures in the data. In medicine, these techniques can use EHR data to classify patients into those that have diabetes and those that do not have it or to cluster diabetic patients into different diabetic subtypes.

In classical machine learning algorithms, the focus is on converting input data into features for supervised and unsupervised learning (feature extraction or feature engineering). There is currently great interest in "deep learning" a subfield that uses massive amounts of labeled data and the power of modern computers to automatically generate useful features and classification models to obtain previously unachievable levels of accuracy.

[2] https://www.youtube.com/watch?v=t-aiKlIc6uk

[3] Braunstein, ML, The Computer in a Family Practice Center: A "Public" Utility for Patient Care, Teaching and Research, *Medical Data Processing*, pp 761–68, Laudet, M, Anderson, J, and Begon, F, Editors. London, Taylor and Francis, 1976.

[4] https://www.youtube.com/watch?v=a65uwr_O7mM   https://www.youtube.com/watch?v=ppkg4mQIgXw

Fig. 1.1 HIS, the first hospital information system. (© 1960 Akron Children's Hospital. All Rights Reserved)

Fig. 1.2 Both new care models and the role of EMRs in them have both been a subject of interest for many decades. Here in late 1975, Congressman Paul G Rogers (D FL) chair of the House of Representatives Subcommittee on Health and the Environment (middle), visits the very innovative Family Medicine Clinic at the Medical University of SC to see one of the first family medicine residency programs. As part of the visit, Department Chair, Dr. Hiram Curry (left), and I showed him the role of the clinic's very early EMR in scheduling, medical records, billing, pharmacy and patient education. The system also provided support for monitoring patient compliance with their prescriptions, population health and utilization review
Source: AAFP Reporter Vol III No 1 January 1976

1.3   Health Information Exchange

Sharing of health data also has a long history dating back at least to the 1980s. Historically a sole purpose infrastructure for sharing was envisioned and many were implemented at a health system, community, regional or state basis. While this infrastructure has had many names over the years, it is usually referred to now as a Health Information Exchange (HIE) or a Health Information Network (HIN). A 2011 paper by Gilad Kuperman describes the basic rationale for HIE's as addressing "key healthcare problems that 'siloed' EHRs do not solve". He goes on to give examples of "problems that could only be addressed by interoperability including support for the patient across transitions of care, the ability to perform longitudinal analyses of care, and public-health needs."[5]

Kuperman also lists some of the technical challenges to HIE creation and sustainability: "the need for standards to represent clinical data, the need to identify a patient consistently as they moved among different providers and a framework to assure the patient's privacy." He goes on to say that there are "also questions of who should play a leadership role to address these issues and the kinds of organizational models that could best support interoperability."

Creating workable and sustainable HIEs in the US was one of the goals of the Office of the National Coordinator for Health IT (ONC) the federal agency that created and manages the HITECH program. ONC provided funding to the states for establishing their HIEs and it also funded the Nationwide Health Information Network (NHIN) Prototype Architecture initiative. The NHIN was envisioned as a 'network of networks' connecting the various regional and state health data exchanges. For both technical and political reasons, the NHIN did not require a national patient identifier or a large-scale centralized operation.

ONC then funded the NHIN Trial Implementations project that eventually involved 20 participating organizations. This effort led to CONNECT and open source gateway for secure HIE. CONNECT was so complex and HIE financial resources were so limited that it was not widely adopted.

In 2010, ONC initiated the NHIN Direct project (Direct). We'll discuss this more slowly and in more detail later but, for those who cannot wait, special Direct servers provide email-based capabilities that are tailored to the special needs of health information sharing. Unlike CONNECT, Direct is 'lightweight' and much easier to implement. Typical Direct use cases[6] involve a primary care physician (PCP) sending a patient summary to a specialist as part of a referral, a care provider sending a visit summary to a patient or the transmission of results from a laboratory to an EHR.

Such a highly distributed exchange introduces the trust problem. Trust is essentially knowing for certain who you are dealing with in an electronic network such as the Internet. It is relatively easy within a health system where everyone is employed

---

[5] https://www.ncbi.nlm.nih.gov/pmc/articles/PMC3168299/

[6] A written description of how a user will use a software system to accomplish a task or goal.

and providers are credentialed. When HIE is extended further – to community providers not employed by or formerly affiliated with a health system and particularly to patients in their environment – trust becomes far more challenging.

DirectTrust was created by Dr. David Kibbe to help solve the trust problem for Direct on a national level. The details are beyond the scope of a non-technical book but, given the focus on FHIR, it is interesting that, in August, 2017, FHIR inventor Grahame Grieve and several members of the DirectTrust Policy Committee posted a white paper that suggests that FHIR resources could be 'pushed' (sent) in Direct Messages and DirectTrust certificates could be used with the FHIR RESTful Application Programming Interface (API) to establish trust.[7] We will define APIs later but this development illustrates the possibility of entirely new means of health information exchange that, unlike the earlier efforts, do not require a substantial sole purpose infrastructure. Given HIE's long standing problem of establishing a sustainable business model, that we will turn to next, this could be quite significant.

Beyond the impediments we have already discussed, finding a sustainable business model is arguably the most difficult challenge for HIEs in the US. Since 2001 the eHealth Initiative has surveyed HIEs. It's most recent 2014 survey included 125 of 267 (47%) identified organizations (74 community-based, 25 statewide, and 26 healthcare systems). Nearly 85% were operational but only 36% were financially sustainable with no outside support.[8] These may seem like low percentages given how long HIE has been available but they represent significant progress. The report provides a great deal of other interesting information in an easily read presentation format so I suggest it for readers with a deeper interest in this topic. One key point is that, of the HIEs surveyed, 64 have customers for their services in support of the new payment models such as Accountable Care Organizations (ACOs) and 52 support new care delivery models such as the patient-centered medical home (PCMH). We will discuss these new payment and care delivery models later on, but this clearly illustrates the key role that financial incentives play in informatics adoption.

1.4 The Interoperability Challenge

From the 1960s and into the 1980s commercial software was a new industry and the computer systems used in hospitals had limited data storage and processing capabilities. Partially as a result of this, commercial hospital software vendors were new, relatively small organizations that focused on a specific hospital department, such as the clinical laboratory or pharmacy. Each of these so called 'departmental systems' was proprietary and each typically had its own data model. Also, Local Area Networks had yet to appear. This made interoperability, or "data sharing," among

---

[7] https://www.directtrust.org/wp-content/uploads/2017/08/Direct-FHIR-Whitepaper_vers_1.4_08102017.pdf

[8] https://www.ehidc.org/sites/default/files/resources/files/2014_eHI_Data_Exchange_Survey_Results_Webinar_Slides_0.pdf

these systems, difficult. As the use of these departmentally specific systems grew, their inability to share data became more problematic. For example, when a new patient was admitted to the hospital, the departmental systems had no automated way to know that, to know their demographic information and even what bed they occupied. Some hospitals actually connected a terminal in the nursing unit to each of the systems so that a user could enter information (often the same information) into all of the systems by going from terminal device to terminal device!

In 1987 this problem eventually led to the formation of the global, non-profit healthcare interoperability standards setting organization, Health Level Seven® (HL7). That led to the development of HL7's hugely successful V2 messaging standard for health data. We will only consider data sharing standards from V2 onward. However, there is a rich history of earlier efforts involving some of the pioneers in health informatics. This is chronicled in a white paper those of you interested in the early development of protocols and data sharing networks in healthcare should read.[9]

Despite its age, V2 is still widely utilized today. According to HL7, 95% of US healthcare organizations use one version or another of V2. There are at least a dozen versions listed on the HL7 site. The standard is based on EDI/X12 technology originally developed by the US transportation industry in the 1960s and then adopted by the grocery and manufacturing industries (all these industries had complex supply chains to manage).

Figure 1.3 is an example of an electronic lab test result being reported using V2. We will explore V2 in more detail in a later chapter but the last two (OBX) lines (segments) provide the actual results. In this example, they are the patient's hematocrit (the ratio of the volume of red blood cells to the total volume of blood) of 45% and their erythrocyte (red blood cell) count of 4.94. Both are accompanied by their respective normal ranges. So, are they normal?[10]

```
MSH|^~\&|LABGL1||DMCRES||199812300100||ORU^R01|LABGL119951022183858 1|P|2.3
   |||NE|NE
PID|||6910828^Y^C8||Newman^Alfred^E||19720812|M||W|25 Centscheap Ave^^
   Whatmeworry^UT^85201^^P||(555)777-6666|(444)677-7777||M||773789090
OBR|||110801^LABGL|387209373^DMCRES|18768-2^CELL COUNTS+DIFFERENTIAL TESTS
   (COMPOSITE)^LN|||199812292128||35^ML|||||||
   IN2973^Schadow^Gunther^^^^MD^UPIN
   |||||||||||^Once|||||CA20837^Spinosa^John^^^^MD^UPIN

OBX|||NM|4544-3^HEMATOCRIT (AUTOMATED)^LN||45||39-49
   ||||F|||199812292128||CA20837
OBX|||NM|789-8^ERYTHROCYTES COUNT (AUTOMATED)^LN||4.94|10*12/mm3
   |4.30-5.90||||F|||199812292128||CA20837
```

Fig. 1.3 A laboratory text result is reported as an HL7 V2 message. It is formatted in the EDI/X12 format that is both concise and cryptic. (Courtesy HL7)

---

[9] http://www.ringholm.com/docs/the_early_history_of_health_level_7_HL7.htm

[10] The answer is yes. The hematocrit of 45 is within the normal range of 39–49 and the erythrocyte count of 4.94 is within the normal range of 4.3–5.9.

This simple example should serve to explain one of the limitations of V2. It was developed at a time when computer memory and storage were both limited and expensive so it is quite intentionally concise, cryptic and, hence, not particularly human readable.

1.5   Exciting and Potentially Transformational Times

Earlier in this section you learned that DirectTrust could be used with the FHIR RESTful Application Programming Interface (API). Non-technical readers may not understand what an API is. In non-technical terms it can be understood as a 'contract' that says to software developers that if you send a request from a 'client' computer (e.g., a phone, tablet, notebook or desktop) to a 'server' (the computer where the information is stored) in the specified format you will always get a response in a specified format or initiate a defined action. The key point is that APIs allow developers to access and even update or delete information and other resources on a remote computer without having to understand the technical details of what's going on in the system they are interacting with. As we will learn, this has a significant positive impact on the development of client software. In essence the entire 'app ecosystem' we all use daily on our smartphones is made possible by the phone's APIs.

Any reader who has searched the Internet for information or for a web site to purchase an item did that using an API Here is an example using Amazon:

https://www.amazon.com/s/ref=nb_sb_noss?url=search-alias%3Daps&field-keyw
ords=size+10+blue+sweater+for+women

You should fairly easily be able to figure out that this hypothetical shopper is looking for a size 10 blue woman's sweater. Try doing this search yourself and you should see a similar string of characters in the field at the top of your browser that displays your query immediately after you click on an icon to initiate the search.

REST is an acronym for **RE**presentational **S**tate **T**ransfer. We mentioned that, using REST, the client and server are independent. The technical term for this is statelessness. Essentially, what this means is that all of the necessary information to handle the request is contained within the request itself. As a result of that, any of the countless thousands of servers that Amazon operates can receive this query and respond to it based entirely on information within the query itself. A moment's reflection should reveal how important that is in managing an environment where queries can arrive at unpredictable times and ask for unpredictable information.

APIs are having a disruptive and transformational impact on many industries. For example, we now take it for granted that we can find and purchase virtually anything with almost no effort. We can conduct most of our banking and financial affairs from anywhere at any time using our phone. We can take a photo of a check and deposit it without visiting the bank or an ATM. This is all done using APIs.

Today, the technologies that support access and sharing of healthcare data are changing rapidly as the number of health-related open APIs and web tools explodes. *The Untapped Potential of APIs in Healthcare*, an article in the Harvard Business Review, begins: "Leaders of most Internet-based businesses have realized the critical importance of using open application programming interfaces (APIs) to expand the reach of their organizations. If the health care industry followed suit, the impact on the quality and cost of care, the patient's experience, and innovation could be enormous."[11]

At present, most would agree that HL7's Fast Healthcare Interoperability Resources or FHIR is the most important healthcare API. FHIR is based on contemporary technologies. Even though it is still under development (the first "normative version" is expected in late 2018), it is achieving dramatic acceptance and adoption. This book focuses on FHIR because it has the potential to solve many of the interoperability problems that have so long impeded informatics from achieving its potential impact on healthcare quality, cost effectiveness and safety. It does this by offering the promise of far more open and facile access to clinical data (with verified identity and the right permissions). It offers the potential that patients will be able to relatively easily aggregate their health data from whatever providers care for them. They manage and take advantage of that data using apps of their choosing and even make decisions about contributing their data for research or other purposes. In short, we are on the verge of a transformation in healthcare powered by the same technologies and forces that have already transformed most other areas of our lives. We conclude this chapter by looking at an actual example.

Demonstration FHIR App: Sepsis Watch

You can see many examples of the transformational potential of an API powered healthcare system by visiting the SMART on FHIR App Gallery.[12] FHIR can be used as the basis for turning electronic medical record systems into open app platforms that are similar in many respects to the app store accessed by the smartphone in your pocket. Much of the attention is on FHIR apps that support physicians in making a more accurate and timely diagnosis and providing more precise, personalized treatment. FHIR also seems poised to solve another longtime challenge in health informatics which is how to give patients access to their complete health record and the tools to use that data to stay well and to more effectively participate in their own care.

---

[11] https://hbr.org/2015/12/the-untapped-potential-of-health-care-apis
[12] https://apps.smarthealthit.org/

Sepsis is a life threatening generalized organ system dysfunction caused by the body's reaction to a serious infection. Its root cause is the immune systems attempt to deal with an infection by releasing biologic agents into the body that in extreme situations result in generalized inflammation of the vital organs such as the brain, heart and kidneys.[13] Sepsis can be very serious with the risk of death as high as 30% and 50% if it is severe. The most extreme form, septic shock, has death rates of up to 80%.[14]

Sepsis can often be detected early if the right laboratory tests are ordered. This is the goal of the Sepsis Watch FHIR app shown in Fig. 1.4. This app helps the physicians order the right tests in a timely manner. You may wonder where such a tool could be usefully applied.

Sepsis is a particularly common problem in intensive care units (ICU), affecting about 37% of admitted patients and is the leading cause of deaths among patients in non-coronary care intensive care units. The annual cost of hospital care for US patients with sepsis is $14 billion.[15]

Given these statistics, the tele-ICU (eICU) at Emory was felt to be the ideal environment in which to pilot real-time analytics for sepsis.[16] It is also ideally suited to test and refine the various steps involved in the predictive analytics pipeline (the series of steps needed to make a prediction).

Emory Healthcare's eICU currently monitors over 100 patients in five hospitals. The nurses and physicians in the eICU (or 'command center') partner with their counterparts at the bedside to provide an extra layer of monitoring. The eICU provides high-level surveillance using multiscale data acquired from the EHR, near-real time views of physiologic waveforms, and trended data derived from those waveforms. One of the major roles of eICU staff and providers is to respond to alerts and convey concerning trends in data for the bedside providers to act upon. Putting primary responsibility on the eICU to be attentive to and respond to alerts can allow bedside providers to focus their time and resources on patient care for the "right" patient. Furthermore, it can minimize 'alert fatigue,' a common phenomenon where busy clinicians fail to consider advice from information systems.

---

[13] https://www.ncbi.nlm.nih.gov/pubmed/16424713

[14] https://www.ncbi.nlm.nih.gov/pmc/articles/PMC3916382/

[15] https://www.ncbi.nlm.nih.gov/pmc/articles/PMC3916382/

[16] Buchman TG, Coopersmith CM, Meissen HW, Grabenkort WR, Bakshi V, Hiddleson CA, Gregg SR. Innovative interdisciplinary strategies to address the intensivist shortage. *Critical care medicine*. 2017 Feb 1;45(2):298–304.

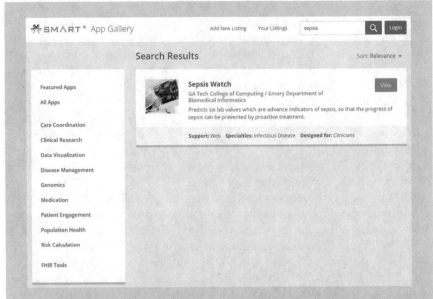

Fig. 1.4 A video demonstration of the Sepsis Watch app is posted on the SMART on FHIR site. The app was developed by a team of Georgia Tech students taking the author's CS6440 graduate seminar and mentored by Dr. Shemim Namati at Emory School of Medicine's Department of Biomedical Informatics. It predicts six lab values that can be indicators of the onset of sepsis.

1.6   A Pivotal Point

The field of health informatics dates back to the early 1960s when computers were first used to store health records. Longstanding interoperability challenges – both technical and non-technical – have impeded the flow of information and progress ever since. We are now at a pivotal point where the building blocks of a more robust and usable technology infrastructure are being developed by HL7 and adopted through the efforts of the government, the health IT and other industries, payers, health care providers, and patients. This book was written in large part to tell that story.

Chapter 2
The US Healthcare System

2.1 Introduction

This chapter briefly describes the US healthcare system and some of the most important of its many problems. This is a complex topic that I cannot adequately cover in a short, introductory book, so I have provided a number of suggested supplemental readings. For a very complete and detailed discussion of the topics raised here (and others) I suggest the Institute of Medicine[1] publication *The Healthcare Imperative: Lowering Costs and Improving Outcomes: Workshop Series Summary* which is available for purchase or free download.[2]

Some readers may wonder why I devote an entire chapter to a topic filled with structural, policy, economic and even political issues in what is, after all, a health informatics book. Based on my prior experience, I know that many readers may have little background in the US healthcare delivery system.

My distinguished former Georgia Tech industrial engineering colleague, Dr. William Rouse, describes US healthcare as a complex adaptive system. Paraphrasing him, such a system is (a) nonlinear and dynamic and does not inherently reach fixed-equilibrium point so it may appear to be random or chaotic and (b) composed of independent agents whose behavior is based on physical, psychological, or social rules rather than the demands of system dynamics.

Because agents' needs or desires are not homogeneous, their goals and behaviors are likely to conflict. In response to these conflicts, agents tend to adapt to each other's behaviors. Agents are also intelligent so, as they experiment and gain experience, agents learn and change their behaviors accordingly. Thus overall system behavior inherently changes over time and may range from valuable innovations to unfortunate accidents. An article in the January 11, 2006 *NY Times* provides a clear

---

[1] The Institute of Medicine (IOM) was a part of the National Academies of Sciences (NAS). In 2016, NAS renamed its divisions and the IOM is now the National Academy of Medicine (NAM – https://nam.edu/) but the IOM name appears on older publications so I continue to use it for those.

[2] https://www.nap.edu/download/12750

© Springer International Publishing AG, part of Springer Nature 2018
M. L. Braunstein, *Health Informatics on FHIR: How HL7's New API is Transforming Healthcare*, https://doi.org/10.1007/978-3-319-93414-3_2

example of this – an innovative network of diabetes care clinics that became "victims of the byzantine world of American health care".[3]

Finally, no one is "in charge." Consequently, the behaviors of complex adaptive systems can usually be more easily influenced than controlled.[4]

We cannot adequately deal with this fascinating and insightful view of healthcare here so I recommend the cited paper to those wanting a fuller treatment of it. However, it seems clear that we do have such a healthcare system and these dynamics have played out over the past few decades with patients, hospitals, providers, employers, payers and the government being the main agents. In what follows we will see that the results have often been quite different from what was hoped and from what most would have desired. It is far too easy to single out one class of agents to assign blame but a more thoughtful and nuanced view would recognize that, for the most part, the system itself creates inevitable conflicts that are extremely difficult to resolve fairly.

Health informatics, the focus of this book, is the application of information technology to that delivery system. For that reason, I believe that some basic grounding in the system is required to appreciate the key roles health informatics can play in solving its problems.

For example, informatics can help avoid medical errors that often arise from inadequate information or the presence of more information than a human being can process. It can also help identify and reduce waste due to unnecessary or duplicative services by offering advice to providers as they write new orders. It is critical to the surveillance of foodborne illnesses and disease outbreaks. Increasingly data mining and many analytic techniques are helping to find new medicines, new treatment protocols, and new methods for earlier diagnosis. These are just a few of the potential informatics benefits to healthcare delivery, but each of them depends to one degree or another on the ability to share digital data.

It should become clear as we proceed that, while medical technology is essential for the treatment of life threatening diseases and other situations such as major trauma, most healthcare and the substantial majority of healthcare costs involve the management of the major chronic diseases (e.g., hypertension, diabetes and chronic heart, lung and kidney diseases). These conditions are often, at least in part, behavioral in their cause. Their successful treatment requires changes in those behaviors, and patients adhering with their physician's recommendation. Since this isn't typically achieved using traditional methods of care delivery, these diseases are often not well controlled, leading to expensive complications that, in our highly specialized healthcare system, are managed by multiple providers who often fail or, even today, don't have a practical means to share data. Addressing these realities is the context behind a number of the most important use cases for health informatics.

All of this points to the key role that interoperability, the meaningful sharing of health information, plays in health informatics. Interoperability is important from

---

[3] https://www.nytimes.com/2006/01/11/nyregion/nyregionspecial5/11diabetes.html

[4] https://www.nae.edu/19582/Bridge/EngineeringandtheHealthCareDeliverySystem/HealthCareasaComplexAdaptiveSystemImplicationsforDesignandManagement.aspx

routine care of a single diabetic patient to research spanning the care of tens of thousands of similar patients. As a result, now that most hospitals and the majority of providers have adopted electronic records, interoperability is increasingly the focus of healthcare policy and informatics standards development. Most of this book will explore the challenges of interoperability. This chapter will provide important background for that discussion.

2.2 High Costs, Mediocre Results

Americans often hear that we have the world's best healthcare system. For over 10 years, Dr. Brent James was Chief Quality Officer at Intermountain Healthcare, which is widely regarded as one of the premier healthcare delivery systems in the US. His data (Fig. 2.1) shows that, compared to three other advanced countries, the US has the lowest mortality rates for high technology medicine – the kind of care often needed to survive a serious heart attack or major trauma. This care is also, of course, extremely expensive and is likely to only grow in cost given the inexorable advances in high technology medicine.

Lower mortality is the positive part of the story. The Commonwealth Fund is a private foundation that promotes a high performing health care system, particularly for our most vulnerable citizens. Figure 2.2 from its international research program shows that, while the US spends around twice as much per capita ($8,508) on

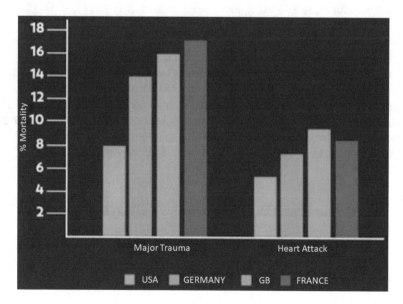

Fig. 2.1 The US healthcare system delivers the lowest mortality rates for life threatening problems such as major trauma or a heart attack which require what Dr. Brent James calls 'rescue care'. (Courtesy Dr. Brent James)

COUNTRY RANKINGS
Top 2*
Middle
Bottom 2*

	AUS	CAN	FRA	GER	NETH	NZ	NOR	SWE	SWIZ	UK	US
OVERALL RANKING (2013)	4	10	9	5	5	7	7	3	2	1	11
Quality Care	2	9	8	7	5	4	11	10	3	1	5
Effective Care	4	7	9	6	5	2	11	10	8	1	3
Safe Care	3	10	2	6	7	9	11	5	4	1	7
Coordinated Care	4	8	9	10	5	2	7	11	3	1	6
Patient-Centered Care	5	8	10	7	3	6	11	9	2	1	4
Access	8	9	11	2	4	7	6	4	2	1	9
Cost-Related Problem	9	5	10	4	8	6	3	1	7	1	11
Timeliness of Care	6	11	10	4	2	7	8	9	1	3	5
Efficiency	4	10	8	9	7	3	4	2	6	1	11
Equity	5	9	7	4	8	10	6	1	2	2	11
Healthy Lives	4	8	1	7	5	9	6	2	3	10	11
Health Expenditures/Capita, 2011**	$3,800	$4,522	$4,118	$4,495	$5,099	$3,182	$5,669	$3,925	$5,643	$3,405	$8,508

Fig. 2.2 Data from the Commonwealth Fund shows that the US spends approximately twice as much as ten other advanced industrialized countries but ranks poorly (11th in four metrics and overall) with respect to care access, efficiency, equity and a healthy citizenry. (Courtesy Commonwealth Fund)

healthcare as the average of ten other advanced, industrialized nations ($4,385), it ranks at the bottom on access, efficiency and equity of care. US citizens also rank at the bottom with respect to living long, healthy, productive lives and the US ranks last overall on these quality measures.[5]

We will now briefly discuss some of the reasons why the US performs so poorly despite spending so much. The reasons for high US healthcare spending are covered well in a white paper from The Bipartisan Policy Center, *What Is Driving U.S. Health Care Spending?* The paper unsurprisingly concludes that "The drivers of health care cost growth are complex and multi-faceted. Just as no single driver is responsible for our high and rising health care costs, no single policy solution will be adequate to meet this challenge."[6] We will only be able to consider a few of these drivers in this book so this paper would be good supplemental material on the topic.

2.3   The Uninsured Can Raise Costs by Delaying Treatment

Lack of health insurance can increase healthcare costs. Figure 2.3 shows that other than the US, virtually all developed nations have universal health care that covers everyone.[7] The Affordable Care Act's major coverage provisions went into effect in

---

[5] http://www.commonwealthfund.org/publications/fund-reports/2014/jun/mirror-mirror

[6] http://bipartisanpolicy.org/wp-content/uploads/sites/default/files/BPC%20Health%20Care%20 Cost%20Drivers%20Brief%20Sept%202012.pdf

[7] https://commons.wikimedia.org/wiki/File:Universal_health_care.svg

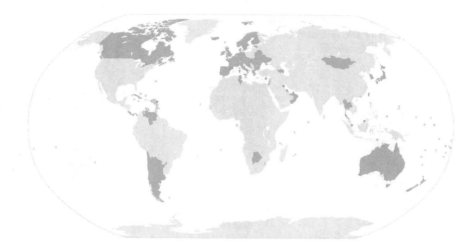

Fig. 2.3 Other than the US, virtually all developed nations have universal health care that covers everyone. (Courtesy Commonwealth Fund)

January 2014 and led to significant coverage gains but, as of the end of 2016, estimates were that around 30 million Americans were still uninsured and its future is far from certain.

The National Academy of Medicine (up through 2016 the Institute of Medicine or IOM) is the healthcare division of the National Academies of Sciences founded in 1863 under President Lincoln. According to the IOM, the uninsured on average use less health care than do insured persons and members of fully insured families. This "lost" utilization may be hidden from view, but it can prove costly in terms of subsequent ill health, disability, and premature death.

The Kaiser Family Foundation (KFF) is a non-profit organization focusing on national health issues, as well as the U.S. role in global health policy. Figure 2.4 presents 2013 KFF data showing that uninsured people receive substantially less healthcare than the insured.

Figure 2.5 presents 2015 KFF data showing that the uninsured are likely to have no regular source of care, they postpone or do without needed care, and postpone or do not obtain needed prescriptions due to cost.

KFF goes on to say that, insured and uninsured people who are injured or newly diagnosed with a chronic condition such as diabetes or hypertension receive similar plans for follow-up care from their physician. However, people without health coverage are less likely to obtain all the recommended services than those with coverage. Finally, because the uninsured are less likely to have regular outpatient care, they are more likely to end up with expensive hospitalizations for avoidable health problems and to experience declines in their overall health. When hospitalized, uninsured people receive fewer diagnostic and therapeutic services and have higher mortality rates than those with insurance.[8]

---

[8] https://www.kff.org/report-section/the-uninsured-a-primer-2013-4-how-does-lack-of-insurance-affect-access-to-health-care/

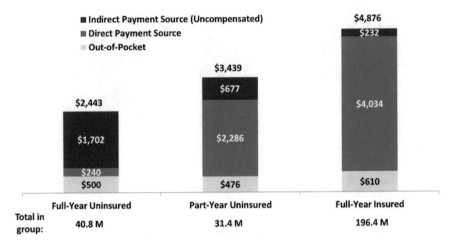

Fig. 2.4 2013 Kaiser Family Foundation data show that the uninsured receive substantially less care, much of it uncompensated, which arguably raises costs for everyone else. (Courtesy Kaiser Family Foundation)

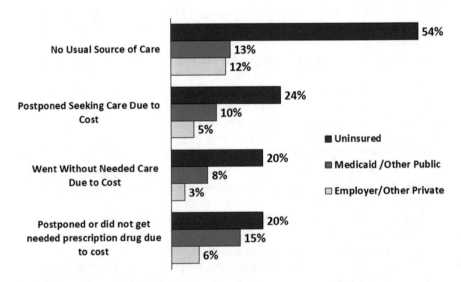

Fig. 2.5 2015 Kaiser Family Foundation data shows that the uninsured are likely to have no regular source of care, to postpone or do without needed care, and to postpone or not get needed prescriptions due to cost. These are all factors that increase the likelihood of far more expensive complications of chronic diseases. (Courtesy Kaiser Family Foundation)

2.4   The Payment Model

Prior to World War II, US patients generally paid for their own healthcare and most could afford it since it was relatively inexpensive. This began to change with the advent of high tech medicine in the 1970s. We tend to associate the use of technology with increased efficiency and lower costs. However, The Bipartisan Policy

Center paper we cited earlier points out that "Advances in medical technology are a major contributor to improving health and increasing longevity, but unnecessary utilization of new technology – especially where a less costly treatment would be equally effective – drives health care spending."

In his 2008 testimony to Congress, Peter Orszag, Director of the Congressional Budget Office, said that "Most analysts agree that the most important factor driving the long-term growth of health care costs has been the emergence, adoption, and widespread diffusion of new medical technologies and services by the U.S. health care system."[9]

Employer sponsored healthcare can be viewed as a war time 'accident'. During World War II there were extreme labor shortages and wages, salaries and "bonuses, additional compensation, gifts, commissions, fees" were frozen but there was an exemption for "insurance and pension benefits". To attract and retain works companies rushed to add health insurance to their employee benefits.[10]

After the war, the US economy experienced explosive growth so employers were still competing for workers. Healthcare insurance was an attractive, inexpensive employee benefit. In the early 1960's the federal government created two major new health insurance programs: Medicare for citizens over age 65 and Medicaid primarily for the poor and certain disabled citizens. Figure 2.6 is from an excellent interactive tool on the California Healthcare Foundation site that illustrates the complex mixture

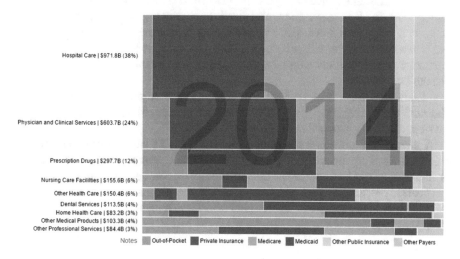

Fig. 2.6 This graphic is taken from an excellent interactive tool from the California Healthcare Foundation site that illustrates the complexity of the US healthcare payment and delivery system by tracking the changes in payments and payment sources from 1960 to 2016. (Courtesy California Healthcare Foundation)

---

[9] https://www.cbo.gov/sites/default/files/cbofiles/ftpdocs/89xx/doc8948/01-31-healthtestimony.pdf

[10] http://www.chicagotribune.com/news/opinion/commentary/ct-obamacare-health-care-employers-20170224-story.html

of payment sources and the variations in their contributions with the kind of care during the period from 1960 to 2016.[11] According to The Bipartisan Policy Center, "Our complex system of payment and delivery leads to increased paperwork and the need for greater administrative resources, raising provider and payer costs." [12]

Until quite recently, most payments for healthcare services were 'pay for procedures' (also called fee-for-service) where, in simple terms, individual providers and hospitals are paid based on how much care they delivered. According to The Bipartisan Policy Center paper "Reimbursement under the fee-for-service (FFS) model generates a strong incentive to perform a high volume of tests and services, regardless of whether those services improve quality or contribute to a broader effort to manage care."

We have touched on a few ways in which the complex and unique US payment model for healthcare drives higher costs. We have also suggested that a major part of these costs could be wasteful.

2.5   Alternate Payment Models

There are other models for payment besides pay-for-procedure. One notable example is Health Maintenance Organizations or HMOs, such as Kaiser Permanente and Intermountain Healthcare. These organizations contract with employers to provide care at a fixed annual cost per employee. This provides an incentive to engineer effective care models that produce superior outcomes at lower cost. Given their financial model, it is not surprising that HMOs were among the early adopters of health informatics.

Another alternate model is Medicare. Since 1982, Medicare has provided a fixed hospital reimbursement for each admission, independent of the actual cost, using a system called Diagnosis-Related Groups or DRGs.

Under DRGs a hospital admitting a patient needing a surgical procedure would receive a fixed payment that might vary based on that patient's medical complexity. If it is able to produce acceptable outcomes at lower cost because it has engineered a superior process, it could make more money. If a patient has complications and stays longer than anticipated, the hospital could lose money.

2.6   Wasteful Healthcare Spending

We will now discuss wasteful spending. I should say at the outset that this is an understandably contentious issue. As you can see in Fig. 2.7, the IOM has identified sources of waste and says that, in aggregate, they account for around 30% of all US healthcare spending, a ratio that has been presented at numerous meetings and in many papers and has remained unchanged for decades.

---

[11] https://www.chcf.org/publication/us-health-care-spending-who-pays/

Fig. 2.7 The IOM shows the sources of what it deems wasteful US healthcare spending. The definition of waste is, of course, subject to argument but, in total IOM says it accounts for some 30% of overall spending, a percentage that has remained unchanged for decades. (Courtesy IOM)

The widely respected journal *Health Affairs* breaks wasteful spending into multiple categories;

First, are failures of care delivery; this refers to poor execution or failure to adopt best practices, such as for effective preventive care or for patient safety.

Second are failures of care coordination; this means fragmented and disjointed care at key times such as when patients move from one care setting to another, called Transitions of Care.

Third is overtreatment; care that ignores scientific evidence: This complex area includes 'defensive medicine' done to avoid lawsuits as well as the use of higher-priced services that have negligible or no health benefits over less-expensive alternatives. This is another major use case for informatics, specifically clinical decision support. You should read the *Health Affairs Policy Brief* for a more detailed discussion.[12]

Fourth is administrative complexity; this is excess spending because payers or regulatory agencies create inefficient or flawed rules and overly bureaucratic procedures. The multiplicity of insurance plans and government programs here in the US almost certainly adds to these costs.

Fifth is pricing failure; this is when the price of a service exceeds that found in a properly functioning market, which would be equal to the actual cost of production, plus a reasonable profit. For example, here in the US, pharmaceutical costs are significantly higher than in other countries. Surprisingly Medicare is prohibited by law from negotiating medication prices with pharmaceutical companies.

The last example of wasteful spending involves fraud and abuse; In addition to the cost of the fraud itself, this includes the cost of additional inspections and regulations to catch wrongdoing.

---

[12] http://healthaffairs.org/healthpolicybriefs/brief_pdfs/healthpolicybrief_82.pdf

2.7   Chronic Disease Drives Most Costs

Earlier we emphasized that the US excels in high technology care, but our overall healthcare as well as its efficiency, access, equity and safety results are poor. In addition to wasteful spending generally, our system suffers because of the high percentage of healthcare costs arising from incurable, lifelong chronic diseases that our healthcare system is not designed or incented to manage well. Patients with multiple chronic diseases account disproportionately for healthcare costs, particularly among those patients aged 65 and above.

Here is the opening paragraph of the excellent 2004 paper *The Growing Burden of Chronic Disease in America* published in Public Health Reports.

> In 2000, approximately 125 million Americans (45% of the population) had chronic conditions and 61 million (21% of the population) had multiple chronic conditions. The number of people with chronic conditions is projected to increase steadily for the next 30 years. While current health care financing and delivery systems were designed primarily to treat acute conditions, 78% of health spending is devoted to people with chronic conditions. Quality medical care for people with chronic conditions requires a new orientation toward prevention of chronic disease and provision of ongoing care and care management to maintain their health status and functioning. Specific focus should be applied to people with multiple chronic conditions.

The paper goes on to describe the care of those patients, saying "while the average Medicare beneficiary sees between six and seven different physicians, beneficiaries with five or more chronic conditions see almost 14 different physicians in a year and average 37 physician visits annually. People with five or more chronic conditions fill almost 50 prescriptions in a year."[13] This stunning fragmentation of care for these people treated sporadically by so many specialists drives health care spending upward and is a central reason to explore and improve health data interoperability.

2.8   Alternate Care Models: Health Maintenance Organizations (HMOs)

Are there examples of a better approach? Yes. Earlier we discussed health maintenance organizations. We will now look at the oldest of those in more detail.

The industrialist Henry J. Kaiser and physician Sidney R. Garfield founded Kaiser Permanente in 1945. Edgar Kaiser famously described the company's economic model as "the less Kaiser does for patients the more money it makes". This is because Kaiser offers a prepaid plan to employers. Kaiser agrees to provide all needed care for a fixed amount per year.

---

[13] https://www.jstor.org/stable/20056677?seq=1#page_scan_tab_contents

Kaiser's history dates to 1933 when Garfield opened the twelve bed Contractors General Hospital to treat construction workers building the Los Angeles Aqueduct in the Mojave Desert. The hospital was in a precarious financial state fueled by Garfield's desire to treat all patients regardless of their ability to pay. Harold Hatch, an insurance agent, proposed that the insurance companies pay the hospital a total amount, in advance, for each worker covered. The financial relationship between the insurance companies and the hospital was efficient and it allowed Garfield to focus on a new idea: preventative health care. This is the first key take away. **When the healthcare economic model provides the right incentives, the focus shifts to preventive care in order to avoid costs.**

Kaiser is also a leader in care coordination and population health, two key strategies for managing chronic disease. To support these strategies, Kaiser has also been a leader in the use of information technology.[14]

The Healthcare Information and Management Systems Society (HIMSS) is a large not-for-profit organization formed in 1961 by Georgia Tech industrial engineering professor, Dr. Harold Smalley, and his former student, Ed Gerner.[15] HIMSS focuses on better health through information technology. Its annual meeting is by far the largest professional event in the field. In 2009 Kaiser received HIMSS' first-ever Stage 7 Award, the organization's highest level of recognition for an environment where paper charts are no longer used. That is the second key take away. **The care models that work best for chronic disease rest on the use of information technology.** Kaiser has done well and today it employees 186,000 staff and 18,600 physicians to serve over 10,000,000 patients in the areas where it operates.

2.9   Alternate Care Models: The Patient Centered Medical Home (PCMH)

A large part of Kaiser's success is because it is concentrated in a few areas where it has the market presence to own and operate all or most of its facilities and employ its physicians. Can smaller care organizations deliver the kind of care that Kaiser provides?

Many feel the Patient Centered Medical Home, or PCMH, can provide a model for improved, more cost-effective community-based care. The concept dates to the late 1960s.[16] Starting around 1970, the author worked in an early example of a clinic

---

[14] Morris F. Collen, MD was one of seven founding partners of The Permanente Medical Group. In 1961 he founded Kaiser Permanente Department of Medical Methods Research (now the Kaiser Permanente Northern California Division of Research). Over the next 50-plus years, this organization was one of the pioneers in the science of using patient data to aid in medical decision-making and to continuously improve care.

[15] http://www.himss.org/sites/himssorg/files/HIMSSorg/Content/files/HistoryHIMSS_January2013.pdf

[16] http://pediatrics.aappublications.org/content/pediatrics/113/Supplement_4/1473.full.pdf

with many of the characteristics now associated with the PCMH where I oversaw the development of one of the first electronic medical record systems. The American College of Physicians, which is the professional organization for internal medicine doctors, advocates the PCMH model. The American Academy of Family Physicians (AAFP), the other large professional association of primary care physicians, also strongly supports the model. This is because the members of both associations are the doctors on the front lines of managing chronic disease.

The ACP describes the PCMH as "a care delivery model whereby patient treatment is coordinated through their primary care physician to ensure they receive the necessary care when and where they need it, in a manner they can understand." It goes on to say that the "objective is to have a centralized setting that facilitates partnerships between individual patients, and their personal physicians, and when appropriate, the patient's family. Care is facilitated by registries, information technology, health information exchange and other means to assure that patients get the indicated care when and where they need and want it in a culturally and linguistically appropriate manner."[17]

Figure 2.8 is a graphic from the AAFP illustrating what it considers the four key elements necessary to achieve a PCMH. It clearly illustrates the importance of health information technology to achieve the goals of this novel care model. This model seeks to use interoperable health information technology to put the primary care physician in a position to coordinate care among multiple providers to avoid unnecessary, redundant or conflicting care.

A 2013 paper titled *Role of Health Information Technologies in the Patient-Centered Medical Home* begins by saying "the Patient-Centered Medical Home (PCMH), requires fulfillment of six standards determined by the National Committee for Quality Assurance to (1) enhance access and continuity, (2) identify and manage patient populations, (3) plan and manage care, (4) provide self-care and community support, (5) track and coordinate care, and (6) measure and improve performance. Information technologies play a vital role in the support of most, if not all, of these standards." It then goes on to discuss many of the potential roles for information technology in the PCMH model as well as the challenges and opportunities for improvement in the technology.[18]

In October, 2013 the Patient-Centered Primary Care Collaborative (PCPCC) released a report titled *Managing Populations, Maximizing Technology: Population Health Management in the Medical Neighborhood.*[19] Figure 2.9 from that report provides a more detailed view of the roles that health information technology can play in this care model and how they align with its key goals.

---

[17] https://www.acponline.org/practice-resources/business/payment/models/pcmh/understanding/what-pcmh

[18] https://www.ncbi.nlm.nih.gov/pmc/articles/PMC3876384/

[19] http://www.pcpcc.org/download/4378/PCPCC%20Population%20Health%20FINAL%20e-Version.pdf?redirect=node/200274

Fig. 2.8 Health information technology is one of four central pillars of the PCMH as identified by the AAFP. (Courtesy AAFP)

PCMH ATTRIBUTE	DEFINITION	SAMPLE HEALTH IT STRATEGIES
Person-centered	A partnership among practitioners, patients, and their families ensures that decisions respect patients' wants, needs, and preferences, and that patients have the education and support they need to make decisions and participate in their own care.	• Care teams use **EHRs** to capture patient needs and medical history, document care plans, as well as information about language, culture, family support, and communication preferences. • **Shared decision-making** and other patient-support tools are made available through patient portals and patient communication.
Comprehensive	A team of care providers is wholly accountable for a patient's physical and mental health care needs, including prevention and wellness, mental health, acute care, and chronic care.	• Care teams used **structured data fields, custom reporting,** and **analytics tools** to track patient outcomes and gaps in care. • **Automated outreach** is sent to patients for gaps in recommended care; and notifications are sent to providers when patients fail to fill prescriptions or miss scheduled immunizations.
Accessible	Patients are able to access services with shorter waiting times, "after hours" care, and/or same day.	• **Telephone or e-mail consultations** are available with clinicians during evenings and weekend hours. • **Patient portals** or **mobile apps** allow online appointment scheduling and email with providers.
Coordinated	Care is organized across all elements of the broader health care system, including specialty care, hospitals, home health care, community services and supports.	• Primary care providers are **alerted** when a patient is admitted or discharged from the hospital. • Interoperable **EHRs** exchange and capture information shared between specialists and primary care providers.
Committed to quality and safety	Clinicians and staff enhance quality improvement through the use of health IT and other tools to ensure that patients and families make informed decisions about their health.	• **Clinical decision support** tools are used to specify order sets for diabetic patients. • **Population health management** tools stratify patients by risk level to determine level of attention from care coordinator.

Fig. 2.9 A more detailed view of the specific roles health information technology can play in achieving the goals of the PCMH. (Courtesy PCPCC)

2.10   Alternate Payment Models: Accountable Care Organizations (ACOs)

Earlier we mentioned that healthcare in the US was traditionally calculated charges based on the amount of care that was delivered. This is usually termed 'pay for procedures' (or 'fee-for-service'). HMOs are an example of an alternate payment model in which there is fixed reimbursement so, if Kaiser spends less than its premium income, it makes a profit. In fact, like most non-profits, it still needs earnings to invest in its business. However, this relies on Kaiser employing all the physicians, owning all the hospitals, and more.

We've now discussed the PCMH as a care model with objectives similar to those found in Kaiser but which can be implemented locally even in a single clinic. Can we create the corresponding *financial model* on a much smaller scale using physicians, hospitals and other community health resources that are already in place? In 2006, Dartmouth professor Dr. Elliott S. Fisher is said to have first used the term "Accountable Care Organization" to describe a proposed solution (although the transcript on page 326 suggests that the commission chair, Glenn M. Hackbarth, may have first used the term).[20] The details are complex but the basic idea is to create a financial model under which existing community providers organize and deliver care in a way that improves the quality and lowers the cost of care by fostering greater accountability on the part of providers for their performance. One of the key provisions of the Obama administration's Affordable Care Act was the creation of ACOs within Medicare.

Medicare initiatives are often adopted by private insurance companies and, by the end of the first quarter of 2017, there were "20 active public and private ACOs across the United States, covering more than 32 million lives" which is around 10% of the US population.[21] Figure 2.10 presents a graphic showing that around half of ACO contracts are now with private insurance companies.[22]

CMS maintains a web site devoted to its efforts to create innovative healthcare delivery and payment models.[23]

2.11   The Role of Health Informatics in ACOs

Dr. Karen Bell was for many years the Chair of the Certification Commission for Health Information Technology (CCHIT), a not-for-profit Health Information Technology certification body with an educational mission that discontinued

---

[20] http://www.medpac.gov/docs/default-source/meeting-materials/november-2006-meeting-transcript.pdf?sfvrsn=0

[21] http://healthaffairs.org/blog/2017/06/28/growth-of-acos-and-alternative-payment-models-in-2017/

[22] https://www.healthaffairs.org/do/10.1377/hblog20170628.060719/full/

[23] https://innovation.cms.gov/index.html

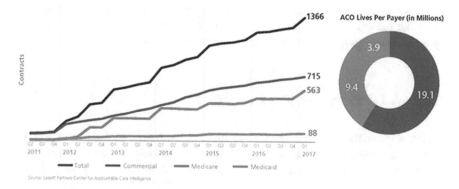

Fig. 2.10 As of early 2017 some 50% of ACO contracts and the majority of patients (covered lives) were covered by commercial health insurance companies. (Used with Permission)

operations in 2014. In a June 2013 post on the Health Affairs blog titled *A Health Information Technology Framework For The Accountable Care Environment,* Dr. Bell said that models such as ACOs must successfully implement seven business processes, all of which rely heavily on the proper use of health informatics for[24]:

- Care Coordination
- Cohort Management
- Patient Relationship Management
- Clinician Engagement
- Financial Management
- Reporting
- Knowledge Management.

To support this she lists these primary requirements of an ACO HIT infrastructure:

- The ability to share health care information between and among various internal and external providers, patients, and their designated caregivers
- Data integration from clinical, administrative, financial, and patient derived sources
- Attention to HIT functions that support patient safety
- Strong privacy and security protections.

Most healthcare organizations have a lot of work to do in order to achieve all of these. Earlier in her post, Dr. Bell said "the promise of accountable care is tempered by a dearth of experience with care-process redesign and culture change and of knowledge about the health information technology (HIT) infrastructure necessary to optimally support health care transformation."

---

[24] http://healthaffairs.org/blog/2013/06/06/a-health-information-technology-framework-for-the-accountable-care-environment/

To summarize, the need for health IT to support new care and payment models is clear, but the industry still has a long way to go to develop and properly use those systems. Next, we turn to some of the key challenges to doing that.

2.12   A Learning Healthcare System

So far, we have focused on the role of health informatics to improve the delivery of care. What about its role in assuring we deliver the right care to every patient every time? This is a major concern of the Academy of Medicine (formerly the IOM). It recognizes that we are entering a new era of medical science that offers the prospect of personalized (precision) health care. It also recognizes that physicians and patients must deal with an increasingly complex array of health care options and decisions, and they will need help to choose wisely. That help, in turn, may require new knowledge about what works and when and for whom it works.

To achieve that, years ago the IOM called for a "Learning Healthcare System"[25] that it describes as a sustainable system that gets the right care to people when they need it, and then captures the results in order to inform further improvement. In simple terms, we need to gather data from care already delivered, aggregate it, and analyze it to learn from the collective results of many care episodes. We need a positive feedback cycle.

2.13   Informatics for a Learning Healthcare System

There are at least three key informatics challenges to overcome in order to achieve a Learning Healthcare System. Providers must adopt digital records. Those systems must be able to share data – interoperability. That shared data must be aggregated, analyzed and presented in a clinically useful and timely manner. Because of the federal government's HITECH program adoption is now a largely solved challenge. Much of the excitement about FHIR rests on its apparent ability to help overcome the interoperability challenge. FHIR apps also offer the prospect of usefully presenting new knowledge gleaned from analysis of past care to busy clinicians in a way that fits well into their clinical workflow and processes.

Here again, economics plays a key role. If providers have an incentive to get it right and do that efficiently they will be interested in creating a Learning Healthcare System because they benefit from it.

---

[25] http://www.nationalacademies.org/hmd/Reports/2007/The-Learning-Healthcare-System-Workshop-Summary.aspx

2.14 Recap

Despite its noble mission healthcare is a business. Historically, here in the United States, healthcare was paid for on a fee-for-service basis which often disincentivizes innovation, cooperation and care coordination. This is particularly antithetical to the treatment of chronic diseases which are rapidly reaching epidemic prevalence so it unsurprisingly that patients with multiple chronic diseases account disproportionately for healthcare costs. The emergence of alternate care and payment models provide economic incentives to solve many of these problems and to help achieve the Institute of Medicine's vision of a continuously improving Learning Healthcare System. The effective and widespread use of informatics is essential to success.

Chapter 3
Health Informatics in the Real World

3.1 Introduction

In this chapter, we begin a discussion of the key informatics tools needed to power a Learning Healthcare System. These include electronic records for both providers and patients as well as information sharing technologies to bind them together. There are two similar terms used to describe these records. Electronic medical records (EMRs) are best thought of as "the standard medical and clinical data gathered in one provider's office".[1] In the early days of clinical information systems, before data sharing was feasible creating the EMR, as defined, was the goal. The newer term is Electronic Health Record (EHR) and it contains a complete record of a patient's care across all providers. Increasingly, with the growth of mHealth (mobile health), it may also contain data contributed by the patients using mobile apps or wearable or other devices in the home.

We begin by discussing the challenges of EHR adoption and the federal programs that largely overcame those challenges in less than a decade.

Finally, we will discuss the many remaining, unsolved challenges in perfecting these systems to make them both usable and useful in the real world of healthcare delivery.

3.2 The Hospital EHR Adoption and Functionality Challenge

As shown in Fig. 3.1, as of 2015 most US hospitals had a 'certified' electronic health record (EHR) system. Certification was the first of three programs developed by the federal government to increase EHR adoption under a law called the Health

---

[1] https://www.healthit.gov/providers-professionals/electronic-medical-records-emr

© Springer International Publishing AG, part of Springer Nature 2018
M. L. Braunstein, *Health Informatics on FHIR: How HL7's New API is Transforming Healthcare*, https://doi.org/10.1007/978-3-319-93414-3_3

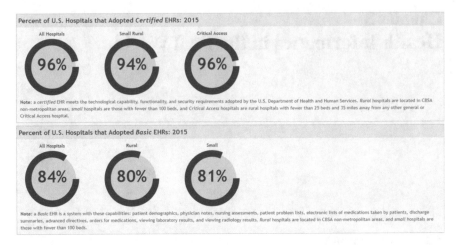

Fig. 3.1 As of 2015 virtually all US hospitals had adopted an EHR but not all of these even met the criteria for a 'basic EHR' which itself is a low bar. (ONC)

Information Technology for Economic and Clinical Health Act (HITECH), a part of the American Recovery and Reinvestment Act of 2009, commonly called the stimulus. We will discuss HITECH later in this chapter but its objective was to provide incentives for EHR adoption by both hospitals and community providers of healthcare. We will first discuss hospitals.

For our purposes of considering EHR functionality, the following condensed list of the Meaningful Use criteria developed by Micky Tripathi, PhD, President and CEO of Massachusetts eHealth Collaborative, is worth considering:

- A core of consistent, structured, clinical content that would be uniform across vendor systems and care settings
- Automated alerts and reminders
- Consistent, robust measurement capabilities
- Data mining capabilities
- Public health reporting
- Interoperability with other systems[2]

It should be clear that these capabilities align with the roles we described earlier for information technology in new care and payment models. That is not a coincidence of course.

While there are many problems with the usability and functionality of EHRs, HITECH was clearly successful in meeting its stated goal of widespread EHR adoption. To illustrate that by putting the current level of adoption into clear perspective, as of 2009 only 1.5% of non-federal hospitals had an EHR.[3] The article cited was published in the prestigious *New England Journal of Medicine* (NEJM) and contains

---

[2] http://bok.ahima.org/doc?oid=105689#.Wd5ZIGhSyUk

[3] http://www.nejm.org/doi/full/10.1056/NEJMsa0900592#t=article

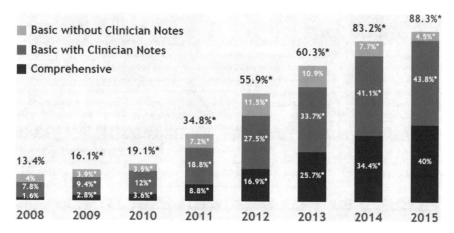

Fig. 3.2 As of 2015 only 40% of hospital installations met ONC's definition of a comprehensive EHR. (ONC)

a working definition of comprehensive and basic EHR systems that eventually found its way into public policy.

Next, we will consider the functionality of the adopted EHRs. Figure 3.2 further breaks down hospital EHR installations based on the comprehensiveness of the EHR's functionality. Most record clinician notes but, as we will now discuss, only 40% use clinical data proactively to try to improve clinical decisions.

Historically, commercial EMRs were developed primarily as a means of assuring that charges and related billing information were properly captured in hospitals. Partially as a result of their historical role, adoption of EHRs by hospitals may be virtually universal today, but the optimal use of those EHRs to achieve a Learning Health System is far from complete. In other words, there is considerable unmet potential of electronic records to improve the quality, safety and efficiency of care delivery as envisioned by the IOM.

To help us explore that functional 'gap', Fig. 3.3 which provides ONC's definition of a Basic EHR without Clinician Notes, a Basic EHR with Clinician Notes and a Comprehensive EHR. To put these definitions into perspective we will consider one characteristic that is only found in a Comprehensive EHR – the ability to actually see image studies rather than just read a text report describing the results and clinical significance of those studies.

The NEJM hospital survey we cited earlier excluded Federal hospitals in large part because the Veteran's Administration (VA), the nation's largest integrated healthcare system, was a pioneer and all their hospitals had EHRs well before the study was conducted. The VA system is enormous and consists of 170 Medical Centers (hospitals) and 1,065 outpatient sites. The VA's Veterans Information Systems and Technology Architecture (VistA) dates back to the 1980s and consists of nearly 180 applications for the VA's clinical, financial, administrative, and infrastructure needs, all integrated into a single, common database. In mid-2017 the VA announced that it will replace VistA with a commercial EHR from Cerner

EHR Functions Required	Basic EHR without Clinician Notes	Basic EHR with Clinician Notes	Comprehensive EHR
Electronic Clinical Information			
Patient demographics	•	•	•
Physician notes		•	•
Nursing Assessments		•	•
Problem lists	•	•	•
Medication lists	•	•	•
Discharge Summaries	•	•	•
Advance directives			•
Computerized Provider Order Entry			
Lab reports			•
Radiology tests			•
Medications	•	•	•
Consultation requests			•
Nursing orders			•
Results Management			
View lab reports	•	•	•
View radiology reports	•	•	•
View radiology images			•
View diagnostic test results	•	•	•
View diagnostic test images			•
View consultant reports			•
Decision Support			
Clinical guidelines			•
Clinical reminders			•
Drug allergy results			•
Drug-drug interactions			•
Drug-lab interactions			•
Drug dosing support			•

Fig. 3.3 ONC's functional description of basic EHRs without and with clinician notes and a comprehensive EHR. Note that only the comprehensive EHR provides clinical decision support, a key part of a Learning Health System. (ONC)

Corporation, the same EHR selected earlier by the US Military Health System. One of the key factors was assuring interoperability between those two health systems. As of this writing the contract has not been signed and the decision to replace VistA is itself controversial.[4]

Figure 3.4 is an example of one of the interoperability capabilities VistA provides. It shows that physicians can retrieve *actual images* from anywhere in the huge nationwide system. This capability is only found in what ONC defines as a comprehensive EHR and so it is still not widely available elsewhere in US healthcare and is particularly uncommon among disparate health systems where the image might be stored in one proprietary system and the physician desiring to see that image is using another proprietary system.

Of more clinical significance is the ability of an EHR to help increase patient safety and avoid medical errors. Here, only a comprehensive EHR has the requisite capabilities to provide reminders and decision support. Even ONC's definition doesn't include many of the more advanced forms of clinical decision support that

---

[4] http://www.healthcareitnews.com/news/results-are-users-say-va-officials-should-keep-vista-ehr-not-sign-cerner

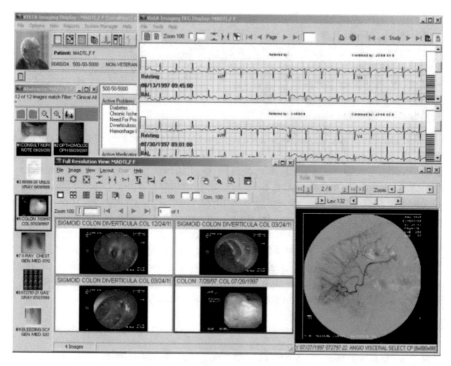

Fig. 3.4 The VA's VistA EHR, despite being quite old, provides the ability to retrieve images across the entire nationwide system, something that is still not widely available across health systems elsewhere in US healthcare. (VA)

are needed, particularly as we move into an era of personalized (or precision) medicine where each individual patient is treated based on their particular and possibly unique clinical, genomic and other factors. As we proceed and learn more about FHIR and SMART on FHIR apps it should be clearer how this new technology may provide a vehicle for adding these important capabilities to EHRs that currently don't have them.

3.3   The Similar Provider EHR Challenges

As with hospital EHR adoption, in less than a decade, we went from 4% of ambulatory (outside the hospital setting) providers having a basic EMR system according to an earlier survey published in the NEJM by the same group of researchers[5] to around half having what is defined as a 'basic EHR' system. The ONC defines a basic provider EHR as one that has these capabilities:

---

[5] http://www.nejm.org/doi/full/10.1056/nejmsa0802005#t=article

- Patient demographics
- Patient problem lists
- Electronic lists of medications taken by patients
- Clinician notes
- Orders for prescriptions
- Viewing laboratory results, and
- Viewing imaging results.

How did this transformation toward a more electronic healthcare system occur in less than a decade? Many people, including the author, long argued that only federal intervention would make EMR adoption happen. A core reason for this is economics. Without federal intervention, hospitals and providers would have to invest to go through a difficult, time-consuming and often frustrating transition from paper to electronic records. Moreover, under the pay-for-procedure reimbursement model, to the extent the EMR led to a reduction in unnecessary or duplicative care, it might negatively affect their revenue. Moreover, EMR adoption is an example of 'network benefits'. The more providers that have them, the more valuable they become because of data sharing. In such a situation, it is always hard to find enough early adopters willing to innovate in the hope of some future gain.

3.4   Health Information Technology for Economic and Clinical Health Act (HITECH)

The details of HITECH are complicated and the program is essentially complete so we will only discuss it briefly at a high level. The EHR adoption program consisted of three main elements:

- HIT Certification that defined minimum functional requirements for EHRs and other clinical tools.
- Meaningful Use that defined the minimum use to which hospitals and providers must put their EHR.
- Incentive Payments made to hospitals and providers for adopting a certified EHR and meeting the requirements of Meaningful Use.

These programs are interdependent and do not apply to all providers. Only hospitals and providers that participated in Medicare or Medicaid; adopted a certified EHR; and used it according to the prescriptions in meaningful use were eligible to receive incentive payments in 2011–2015.

Incentive payments came to hospitals and providers through their participation in either Medicare or Medicaid. Providers could be eligible for only one of these payments but Hospitals could be eligible for both. There were annual limits to the amount of the payments and the total amount that a provider could earn depended on when that provider achieved the stages of Meaningful Use. By October 2015, more than 479,000 health care providers received incentive payments.

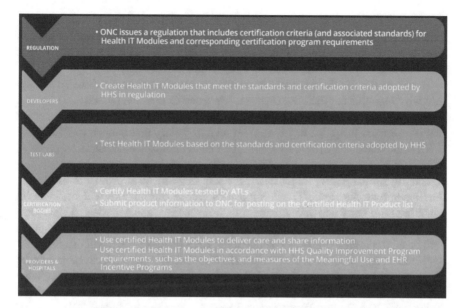

Fig. 3.5 A diagrammatic overview of ONC's HIT certification process begins with the development of regulations that drive system development. The resulting systems are tested by authorized labs and certified by authorized organizations before being put into use by providers and hospitals. (ONC)

3.5   Health IT Certification

The Office of the National Coordinator for Health Information Technology (ONC) administers the HIT certification program. It is voluntary and includes capabilities such as the recording, securing, and interoperable sharing of health information.

Figure 3.5 shows how the program defines the technical requirements for health IT and the process by which commercial systems may become and remain certified. Commercial Authorized Testing Labs do the testing using criteria and test datasets provided by ONC.

Since hospitals and providers must acquire a certified EHR to receive incentive payments it is virtually a business necessity for EHR vendors to be certified. By July 2016, 632 vendors supplied certified health IT to over 337,000 ambulatory primary care and specialist physicians and other providers participating in the Medicare EHR Incentive Program. While, the most widely installed vendor, Epic Systems, supplies certified technology to 30% of providers the large number of vendors and systems clearly exacerbates the interoperability challenge.

As we saw earlier, through program year 2016 of the Medicare EHR Incentive Program, virtually all of the 4,520 non-federal acute care hospitals installed certified EHR systems. While 186 health IT developers supplied systems, Epic, Cerner and MEDITECH each had nearly 1000 installations and over 98% of hospitals had purchased a system from one of the 10 most commonly installed vendors. Despite this vendor concentration, interoperability across health systems remains a problem. The reasons are not limited to technology: too many hospitals and health systems seek to control access to their digital records to increase the likelihood that their patients will obtain all their care from within their system.

However, despite their interest in maintaining perceived competitive advantage through capabilities of their own systems, Epic and Cerner, the two largest vendors, have announced support for FHIR and SMART on FHIR as an EHR-connected platform for FHIR apps. They both offer an app gallery that is open to third party FHIR app developers after they go through each company's certification process. MEDITECH, the third major healthcare enterprise software vendor, is developing something similar. Other major enterprise and large medical providers EHR vendors have done the same. These companies are, or presumably will be, even be soliciting third party app developers for their respective 'app galleries'. This liberation of EHR data for innovative purposes may not be classic interoperability but it could lead to progress in solving some of the EHR challenges we will discuss later.

While an app ecosystem business model like that so common in the smartphone industry may be attractive to these EHR vendors, it is possible they are also responding to pressure from the federal government particularly with respect to patient-facing apps. The Twenty-First Century Cures Act was passed in late 2016. One of its aims was to make EHR data more accessible so it states that open APIs will be necessary to maintain HIT certification starting in late 2017. Specific quotes from the act posted on the SMART on FHIR site require:

> ... that the entity has in place data sharing programs or capabilities based on common data elements through such mechanisms as application programming interfaces without the requirement for vendor-specific interfaces;
>
> [that they] publish application programming interfaces and associated documentation, with respect to health information within such records, for search and indexing, semantic harmonization and vocabulary translation, and user interface applications; and
>
> [and that they] demonstrate to the satisfaction of the Secretary that health information from such records are able to be exchanged, accessed, and used through the use of application programming interfaces without special effort, as authorized under applicable law.[6]

The quote from a speech at HIMSS 2018 by the new CMS administrator, Seema Verma, at the outset of this book suggests that the current administration is fully supportive of these objectives and understands the critical importance of open APIs to achieving them. HIMSS is by far the largest meeting of the health information technology industry each year.

3.6   Meaningful Use

Meaningful Use is a complex federal program announced with three stages. We can only cover the main points here. The central goals of Meaningful Use are to assure the widespread use of certified EHR technology:

---

[6] https://smarthealthit.org/2016/12/21st-century-cures-act-makes-apis-in-ehrs-the-law/

- In a clinically meaningful manner, such as e-prescribing.
- For electronic exchange of health information to improve quality of health care.
- To submit clinical quality and other measures.

The sub-goals for clinical care include:

- Improving care coordination
- Reducing healthcare disparities
- Engaging patients and their families
- Improving population and public health
- And ensuring adequate privacy and security

Note that these goals align well with what experts believe is needed to improve the management of chronic disease. The developers of Meaningful Use were well aware of the mismatch between the focus of our healthcare system and the diseases that drive most costs.

The Meaningful Use program began in 2010 and, starting in 2015, there were reductions in Medicare payments to hospitals and providers that were not meaningful users. The timeline, and stage 3 in particular, are highly political and many argue that the pace of the program is too rapid given the complexity of implementing and properly integrating EHR systems into a healthcare organization.

3.7 Beyond Meaningful Use

The Medicare Access and CHIP Reauthorization Act of 2015 (MACRA) modifies and may eventually replace the Meaningful Use Stage 3 incentive program. It definitely increases the focus on using electronic health record (EHR) data for value-based care.

Under MACRA, a certified EHR must:

- Indicate the data source for measures, activities and objectives to be reported as part of the Medicare value-based payment programs.
- Transmit data from the certified EHR technology or through a data intermediary in the form specified by CMS.
- Allow individual Merit-based Incentive Payment System (MIPS) eligible clinicians and groups to submit data directly from their certified EHR technology in the form specified by CMS.

The Improvement Activities performance category is a part of MIPS that assesses how much a healthcare organization or physician participates in activities that improve their clinical practice. Those activities include:

- Expanded practice access
- Population health management
- Care coordination
- Beneficiary engagement
- Patient safety and practice assessment

- Participation in an alternative payment model
- Achieving health equity
- Integrating behavioral and mental health
- Emergency preparedness and response.

Once again these align well with what experts believe is needed to improve the management of chronic disease.

There is also an Alternative Payment Model (APM) under MACRA's Quality Payment Program under which physicians that qualify can earn a 5% bonus payment if they achieve threshold levels of payments or patients under one of several specified care models for a specific clinical condition such as end stage renal disease or cancer, a care episode such as a joint replacement, or a population.[7]

ONC maintains an informative site for information about both Meaningful Use and MACRA.[8]

As this book was going to the publisher CMS Administrator, Seema Verma, announced some important changes to several of its programs including Meaningful Use which is being renamed "Promoting Interoperability." This part of the April 24, 2018 announcement is particularly relevant to our discussion:

The proposed policies released today begin implementing core pieces of the government-wide MyHealthEData initiative[9] through several steps to strengthen interoperability or the sharing of healthcare data between providers. Specifically, CMS is proposing to overhaul the Medicare and Medicaid Electronic Health Record Incentive Programs (also known as the "Meaningful Use" program) to:

- make the program more flexible and less burdensome,
- emphasize measures that require the exchange of health information between providers and patients, and
- incentivize providers to make it easier for patients to obtain their medical records electronically[10]

While a seemingly increased focus on interoperability appears to be a positive development, the significance of this announcement is not clear as of this writing.

3.8   Physician EHR Satisfaction

Physician satisfaction with their EHR is an extremely important topic and one that will have widespread ramifications for the effort to use health informatics as a tool for establishing a Learning Healthcare System and effectively using the new knowledge it creates.

---

[7] https://qpp.cms.gov/apms/overview

[8] https://www.healthit.gov/topic/meaningful-use-and-macra/meaningful-use-and-macra

[9] https://www.cms.gov/Newsroom/MediaReleaseDatabase/Press-releases/2018-Press-releases-items/2018-03-06.html

[10] https://www.cms.gov/Newsroom/MediaReleaseDatabase/Press-releases/2018-Press-releases-items/2018-04-24.html

Responding "Agree" or "Strongly agree" to the following:

Our electronic health record improves my job satisfaction	35
In our practice, our electronic health record improves the quality of care	61
Our electronic health record requires me to perform tasks that other staff could perform	61
Using an electronic health record enhances patient-doctor communication that is not face-to-face	54
When I am providing clinical care, our electronic health record slows me down	43
Our electronic health record improves my job satisfaction	38
Using an electronic health record interferes with patient-doctor communication during face-to-face clinical care	36
I receive an overwhelming number of electronic messages in this practice	31
Based on my experience to date, I prefer using paper medical records instead of electronic records	18

Fig. 3.6 The 2013 RAND/AMA survey of physician satisfaction included satisfaction with their EHRs. The results paint a mixed picture with physicians agreeing that their EHR improves care quality but expressing various problematic issues. Overall, however, only 18% of physicians would revert to paper. This is Table 7.1 of the posted study which, as shown here, repeats the first question with slightly different results

In 2013, the RAND Corporation and the American Medical Association published what is, though out of date, arguably still the best study to date on overall physician satisfaction, including satisfaction with EHRs. The survey team selected thirty physician practices to achieve a good cross section. Each practice completed a structured questionnaire that included questions about its electronic health record use and capabilities and participation in innovative payment models. The survey team then visited each practice and conducted semi-structured interviews with 220 informants. Finally, they surveyed 656 physicians in the 30 practices, receiving 447 responses (a 68% response rate).

Figure 3.6 presents the key results with respect to satisfaction with the practice's EHR. The majority of physicians believe their EHR improves care quality but using it is inefficient of their time. Interestingly only 18% of the physicians reported that they would revert to paper records if they could.[11]

According to the American Academy of Family Practice (AAFP), a strong advocate of EHRs, physician satisfaction with them may be declining. As our baseline, in the AAFPs 2010 survey, 61% of respondents said they were "satisfied" or "very satisfied" with their EHRs.

AmericanEHR Partners is a new effort supported by 20 professional societies including the AAFP and the American College of Physicians, another large physician group that advocates for EHRs. AmericanEHR has been collecting data since the AAFP survey in 2010, and it claims theirs is the only EHR survey that charts satisfaction over time and allows for the benchmarking of data. To date, AmericanEHR says it has collected over 7,500 responses using the same core survey. They say this

---

[11] https://www.rand.org/pubs/research_reports/RR439.html

allows them to have comprehensive breakdowns by specialty, practice size, EHR type, etc.

The AmericanEHR Partners' 2014 online survey of 940 physicians found that:

- 42% thought their EHR system's ability to improve efficiency was difficult or very difficult.
- 72% thought their EHR system's ability to decrease workload was difficult or very difficult.
- 54% found their EHR system increased their total operating costs.
- 43% said they had yet to overcome the productivity challenges related to their EHR system.

However, 36,318 physicians received invitations, so the response rate was 2.6%. It is certainly possible that dissatisfied physicians were more motivated to express those feelings, a phenomenon known as response bias. I asked AmericanEHR about that and they say that the 2014 results are actually consistent with results that they have been continuing to receive since then.[12]

In any case, it seems clear that physicians find the use of an EHR inefficient and challenging. We next turn to why this is the case.

3.9 EHR Challenges

In 2009, the National Academies of Science (NAS) published *Computational Technology for Effective Health Care: Immediate Steps and Strategic Directions*. Dr. William Stead, the distinguished head of Vanderbilt University's Biomedical Informatics Department, was the co-editor. The report states that:

> IT applications appear designed largely to automate tasks or business processes. They are often designed in ways that simply mimic existing paper-based forms and provide little support for the cognitive tasks of clinicians or the workflow of the people who must actually use the system. Moreover, these applications do not take advantage of human-computer interaction principles, leading to poor designs that can increase the chance of error, add to rather than reduce work, and compound the frustrations of executing required tasks. As a result, these applications sometimes increase workload, and they can introduce new forms of error that are difficult to detect.[13]

This seems congruent with the results of the physician EHR surveys cited earlier. It is also suggestive of the concern physicians often express that their EHR provides a 'one size fits all' approach to charting that does not capture the nuances of actual patient care.

---

[12] http://www.americanehr.com/research/reports/Physicians-Use-of-EHR-Systems-2014.aspx
[13] https://www.nap.edu/catalog/12572/computational-technology-for-effective-health-care-immediate-steps-and-strategic

In fact, a very common approach to electronic charting is a template that, as the report suggests, mimics existing paper forms and largely ignores the potential of a computer to be adaptive to the clinical situation.

The report identified seven 'information intensive' aspects of the IOM's vision of a Learning Healthcare System that current EHRs often fail to provide. They are abbreviated here for legibility:

- Comprehensive data on conditions, treatments, and outcomes;
- Support to help providers and patients integrate patient-specific data where possible and account for any uncertainties that remain;
- Support to help providers integrate evidence-based practice guidelines and research results into daily practice;
- Tools to help providers manage a portfolio of patients and to highlight problems as they arise both for an individual patient and within populations;
- Rapid integration of new instrumentation, biological knowledge, treatment modalities;
- Accommodation of growing heterogeneity of locales for care provision, including home instrumentation for monitoring and treatment, lifestyle integration, and remote assistance; and
- Empowerment of patients and their families in effective management of health care decisions and their implementation.

This is in essence a list of what could be called EHR 'clinical deficiencies'. Some of these same deficiencies are discussed in Fix the EHR!, a well written post on the *Health Care Blog* co-authored by Dr. Robert Wachter, author of the NY Times best-selling book *The Digital Doctor.*[14] Many are of increasing interest because of the new care and payment models we have discussed. However, virtually all current EHRs date from the era of pay-for-procedure and emphasize support of billing and administrative processes. Despite efforts by many vendors to deal with usability issues, their underlying purpose of collecting billable charges often makes them awkward and inefficient tools for managing and documenting patient care. They are complex, large-scale systems that can be difficult, costly and time consuming to update. They are also very expensive, particularly at the health enterprise level where total costs including deployment and training can be in the hundreds of millions of dollars. Because of HITECH, they are now widely installed and unlikely to be replaced soon. How might we 'fix' them in place?

3.10 A Universal Health App Platform

In 2014 the Agency for Healthcare Research and Quality (AHRQ) commissioned JASON, an independent group of elite scientists that advised the United States government, to review the interoperability challenge. JASON produced a report titled *A*

[14] http://thehealthcareblog.com/blog/2018/04/07/how-to-fix-the-electronic-health-record/

Robust Health Data Infrastructure that proposed a solution that hinted at a universal health app platform. It seemed at the time highly unlikely to happen.[15] Such a platform, were it to exist, might well provide a vehicle for expanding EHR functionality.

On December 4th, 2014, HL7 CEO Charles Jaffe, MD announced the formation of the Argonaut Project. The Argonaut web page credits several prior efforts, including a review of the JASON report produced by the JASON Task Force.[16] It describes itself as "a new initiative to accelerate the development and adoption of HL7's Fast Healthcare Interoperability Resources (FHIR), with support from 11 organizations, including EHR vendors Epic and Cerner and health systems Mayo Clinic and Intermountain Healthcare."[17] This widespread membership in Argonaut meant that its approach would become the industry standard for what it describes as "expanded information sharing for electronic health records and other health information technology based on Internet standards and architectural patterns and styles."[18]

Figure 3.7 presents a software architecture proposed by the JASON report that goes on to describe it as follows:

"The top of the architecture, labeled "UI apps" (user interface applications), contains all of the applications that interface with the physical world. In the JASON architecture, a clinician's tablet display interacts with the HIT system through UI apps. The patient's interface to his/her personal health record is through an app, perhaps running on a mobile device. If the results of a diagnostic test are entered automatically into a health record, the software that does so is an UI app. Payers also interface with the HIT system through UI apps. All stakeholders in the system interact with the architecture through applications in the UI apps layer."[19]

As we will discuss in more detail later on, FHIR adoption is now widespread and it became the technologic basis for a universal health app platform.

3.11   Innovative EHR Functionality

In this section we will look at three innovative EHR designs intended to address some of the deficiencies frequently cited by physicians. The first takes advantage of FHIR as an app platform to transform the way an existing commercial EHR operates while the others use machine learning as part of an innovative EHR design.

---

[15] https://www.healthit.gov/sites/default/files/ptp13-700hhs_white.pdf

[16] https://www.healthit.gov/sites/default/files/facas/Joint_HIT_JTF%20Final%20Report%20v2_2014-10-15.pdf

[17] https://www.healthcare-informatics.com/article/argonaut-project-build-jason-s-fhir-recommendations

[18] http://argonautwiki.hl7.org/index.php?title=Main_Page

[19] https://www.healthit.gov/sites/default/files/ptp13-700hhs_white.pdf

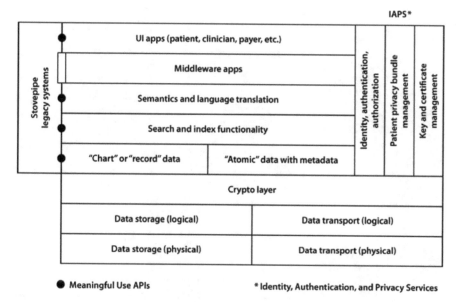

Fig. 3.7 The 2014 AHRQ commissioned report, *A Robust Health Data Infrastructure*, envisioned a new architecture in which patients, providers, payers and other stakeholders can access data through "a software ecosystem, with a diversity of products and apps that fosters innovation and entrepreneurship." (AHRQ)

Geisinger Health System   The Geisinger Health System in central Pennsylvania is widely regarded as one of the nation's highest performing health care organizations. It was an early adopter of the approach recommended by JASON and of FHIR as a technology to implement it. Figure 3.8 presents Geisinger's EnrG I Rheum™ FHIR app that seems to embody many of the principles espoused by JASON. Its user interface is specific to the management of arthritis, and it can display both the physician's notes and data recorded earlier by the patient, saving physician time and promoting patient engagement.

Given this background it is not hard to imagine some future state where the systems we now think of as electronic health records generally serve as a repository of clinical data and an interface from it to the other operational systems within a clinic or hospital. Providers, patients and other end users interact with all that data via tools finely crafted to meet their needs for recording and visualizing data and more usefully support their workflow and clinical processes.

Praxis   Praxis is an innovative commercial approach to more clinically intelligent EHRs that uses machine learning (specifically a neural network) to enable the EHR to learn and adapt to the clinical approach used by specific physicians to care for the problems they commonly see. As shown in Fig. 3.9, once Praxis is trained on a problem, when the physician next sees that problem, it presents their likely note but the physician must validate each clinical concept for the current visit by clicking on

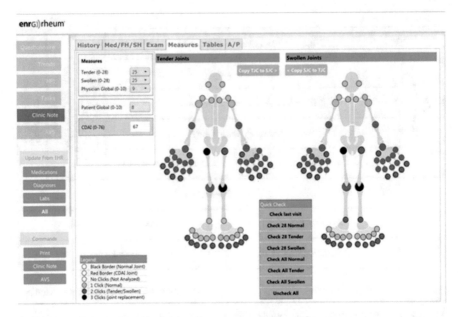

Fig. 3.8 Geisinger's EnrG I Rheum™ FHIR app is an early example of the innovation that JASON predicted would be stimulated by an open health app platform. It seems to seek to support the rheumatologist's mental model as called for by the NAS report cited earlier. (Courtesy Geisinger Health System)

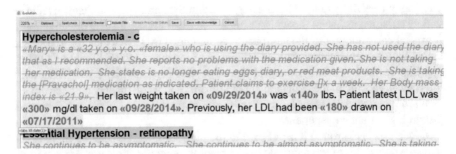

Fig. 3.9 The Praxis EHR uses a neural network to 'learn' how each physician manages the problems they commonly see. Once it is trained (after around 50 instances of a problem) it can predict the care the patient will receive and it generates a grayed out note representing that care. The EHR recognizes fields that would typically vary from visit to visit and these 'Datum' fields are indicated in brackets. Physicians click on grayed out 'clinical concepts' to indicate they are applicable for this visit and on Datum to specify the value of that field. (Courtesy Praxis)

it and Praxis has enough clinical knowledge to recognize the data fields *must* be specified at each visit. The company claims that once the system is trained it actually saves physicians time. The training typically requires 50 visits for a particular clinical problem. The approach seems to work since physicians consistently give Praxis some of the highest ratings for EHR user satisfaction.[20]

---

[20] http://www.aafp.org/fpm/2015/0100/p13.html

LEMR   An interesting 2016 paper from the University of Pittsburgh in the *Journal of Biomedical Informatics* reported an early prototype of what the group called a "Learning Electronic Medical Record System (LEMR).[21] The idea is to train a predictive model to know what clinical information would be viewed as important by a physician in caring for a particular patient at a particular time and automatically display it. The study focused on patients in an intensive care unit (ICU) where, as we discussed earlier, information overload is understood to lead to important clinical events being overlooked. Figure 3.10 is a screen shot from the system. For any given patient, the items highlighted in Panel D, the Highlighted Information Display

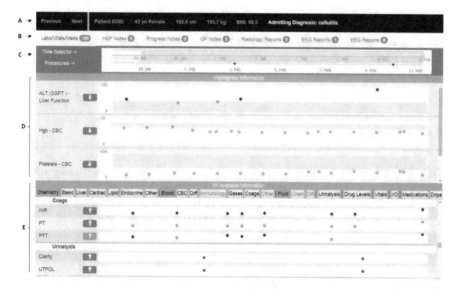

Fig. 3.10   The details of the LEMR information display. Panel A is the patient demographics toolbar, allows the user to move between patients and gives a brief summary of the current patient's demographic information and admitting diagnosis. Panel B contains quick access tabs for navigating among the various types of patient data, including laboratory test results, medication orders, and clinical text reports (e.g., history & physical (H&P) reports, progress notes, and operative procedure notes). Currently the "Labs/Vitals/Meds" tab is selected. This prototype uses times-series plots to display these non-text based clinical items (laboratory test results, vital signs, medication orders, and intake and output data). Panel C is the time range selector, is used to define time ranges of data to display. Below the time-range selector is the procedures axis, which is labeled with the defined times. Black diamonds on this axis represent procedures (surgeries, biopsies, etc.) that the current patient has had. Hovering over a diamond gives more details on that procedure. Panel D, the HID, shows high detail plots of selected clinical items. These plots have a labeled y-axis and blue bands to indicate the normal range. Panel E displays all available results, including those found in the HID, using plots with condensed y-axes. These plots give a notion of trends over time and are arranged by group type (chemistry, lipids, cardiac, etc.). The buttons across the top of this panel list all of the different group types and can be used to jump to a specific type. For both Panel D and Panel E, different colors are used to indicate when a value is within or outside of normal range (blue = below; green = within; red = above; black = no defined normal range). (Courtesy Dr. Gregory Cooper)

(HID), are meant to be the ones that are most clinically relevant for that patient at the current time. The HID can be populated with items by both automatic and manual means. For automatic population, the LEMR uses stored statistical models to predict the probability that each item is relevant given the current patient state. The items that have a predicted probability above a set threshold are displayed. To make manual changes to the contents of the HID, a user clicks on the arrow buttons next to each

The study authors report that physicians were "generally enthusiastic about the LEMR approach and identified advantages, such as adaptation to different specialists and the potential for time savings. Concerns included the feasibility of the implementation and the possible implications of integration into workflow. For instance, some participants worried that over-reliance on highlighted items might cause physicians to miss important details in the remainder of the record. We will see how FHIR now makes it far more feasible to integrate an innovative tool like LEMR into an actual EHR.

3.12 Recap

This chapter concludes Part I of the book. The certification of electronic health record systems and financial incentives offered through the HITECH act have led to rapid and widespread deployment of health IT tools. Far too often these tools have significant usability issues that limit their attractiveness and value to providers. They are also largely disconnected and have therefore not yet met their full potential to impact the challenges we discussed in the prior chapter. FHIR and solutions based on FHIR are starting to enhance the core capabilities of these and other health IT systems and extend their functionalities and usability in ways that could be truly transformational. In Part II we will explore some of those possibilities.

Part II
Beyond Direct Patient Care

Chapter 4
The Empowered Patient

4.1 Introduction

Patient facing health informatics is an increasingly rich and interesting field, particularly because many people now use smartphone apps, wearable and other sensors and mobile devices to monitor and report on their behavior, activity and physiologic parameters. Those new technologies now have a name – mHealth (mobile health) – that will be the topic of a later chapter. In this one, we will address some of the other tools and approaches to patient engagement through informatics. We discuss various examples of personal health records and related tools, the technologies and regulatory incentives that enable them, and the factors that have limited their usefulness and adoption by patients. We also contrast traditional and essentially 'read-only' patient portals with an array of newer interactive tools that provide patients with greater ability to gain insights into their own health information to better manage their care. Some of these are old ideas while others are quite new. Changing economics and the new models of payment and care are increasing interest in mHealth based approaches to patient engagement in their own care as well as connecting patients to their providers on a more continuous basis.

4.2 Personal Health Records

One of the main use cases for interoperability is enabling coordination among the providers taking care of a patient. Today, there is increasing recognition of the importance of other equally compelling use cases for interoperability. Key among them is the ability of patients to be the aggregation point for their health record so that they can make sure all their providers have a complete view of their care. The data in their record can be used by tools and apps to help guide and inform patients using the specifics of their clinical situation.

© Springer International Publishing AG, part of Springer Nature 2018
M. L. Braunstein, *Health Informatics on FHIR: How HL7's New API is Transforming Healthcare*, https://doi.org/10.1007/978-3-319-93414-3_4

It is key that patients be able to contribute data to that record using mHealth devices and apps. The chronic diseases that drive most healthcare costs often have behavioral causes such as a poor diet or lack of physical exercise. Their successful treatment typically involves modifying those behaviors and obtaining patient compliance for their treatment program, often including medications.

A novel and, for many, surprising use case is that patients have access to their physician's notes and the ability to suggest corrections. As we will see, this new interoperability paradigm can have surprising positive benefits.

These are some of the key potential benefits that could arise from the concept of a 'personal health record' or PHR. Its history is fascinating and reveals, once again, that many of the most exciting contemporary developments in health informatics are actually quite old ideas foreseen by visionaries in the field.

The term PHR first appeared in a 1978 paper[1] but it was not practical until around 2000 with the dramatic decline in the cost of computing and the increasing availability of the Internet. In 1994, at the dawn of the Internet, Peter Szolovits, Jon Doyle and William J. Long at MIT's Laboratory for Computer Science, Isaac Kohane at Boston Children's Hospital and Stephen G. Pauker at New England Medical Center envisioned a lifelong patient-centered health information system. In their 'manifesto' they "propose a major shift of primary focus away from information systems based on the hospital, clinic and medical practice, to one based on the individual. The system which they call "Guardian Angel" (GA), integrates over a lifetime all health-related information about an individual (its "subject"), thus providing, at minimum, a comprehensive medical record that is often virtually impossible to reconstruct in a timely manner as the subject moves through life and work assignments."

Their vision went far beyond a passive repository and they listed these ten specific ways Guardian Angel would actively engage patients in their care:

1. engages in data collection, sometimes by interacting with the subject and sometimes by automatic tracking and recording of instruments,
2. monitors the progress of medical conditions and the effects of therapy with respect to expectations, and checks for side effects,
3. interprets facts and medically-related plans and helps explain them to the individual,
4. allows the patient to customize therapy plans within bounds established by care providers, giving the patient "ownership" of his or her therapy,
5. performs "sanity checks" on the appropriateness of diagnostic conclusions and therapeutic plans,
6. contains some understanding of the subjects' preferences, represents these in a broad range of negotiations with other systems, including setting therapeutic guidelines and scheduling appointments, and also uses these in structuring interactions between GA and the patient,

---

[1] "Computerisation of personal health records." Health Visitor. 51 (6): 227. June 1978.

7. interfaces to information systems used by care providers, insurers, researchers, etc., to provide access to personal medical history information as authorized by the individual, and
8. provides patient education functions, including access to general and specialized medical encyclopedias, and explanations of diagnostic findings and therapy plans specific to the individual,
9. implements patient reminding and alerting functions, including reminders of scheduled therapy, medications and appointments, integrating these with personal scheduling tools,
10. provides patient support functions such as contacts with support groups and other patients, queries to pharmaceutical companies, government agencies, etc.[2]

In 2001 Mandl, Szolovits and Kohane (Ken Mandl is also at Boston Children's Hospital) published *Public standards and patients' control: how to keep electronic medical records accessible but private*. The summary points from that prescient article provided a clear roadmap for the future:

Electronic medical record systems should be designed so that they can exchange all their stored data according to public standards
Giving patients control over permissions to view their record – as well as creation, collation, annotation, modification, dissemination, use, and deletion of the record – is key to ensuring patients' access to their own medical information while protecting their privacy
Many existing electronic medical record systems fragment medical records by adopting incompatible means of acquiring, processing, storing, and communicating data
Record systems should be able to accept data (historical, radiological, laboratory, etc.) from multiple sources including physician's offices, hospital computer systems, laboratories, and patients' personal computers
Consumers are managing bank accounts, investments, and purchases on line, and many turn to the web for gathering information about medical conditions; they will expect this level of control to be extended to online medical portfolios[3]

4.3   A Personally Controlled Health Record (PCHR)

The first implementation of what the same group at Boston Children's Hospital described as a free, open source Personally Controlled Health Record (PCHR) may have been the Personal Internetworked Notary and Guardian (PING).[4] They designed PING to meet some key policy requirements and to support population-wide

---

[2] https://groups.csail.mit.edu/medg/projects/ga/manifesto/GAtr.html

[3] http://www.bmj.com/content/322/7281/283

[4] J Am Med Inform Assoc. 2005 Jan-Feb;12(1):47–54.

research. Chief among the policy objectives was that the PCHR exist outside the administrative structures of any particular health care institution. At the time (and this is frequently still true today) personal health record systems are often institution- or company-specific repositories for health information. PING was designed to be an interoperable, patient-controlled health record. It allowed each patient to maintain their own encrypted record in a storage site of their choosing. The developers envisioned that storage sites might be server farms or individual Internet accounts. Access, authentication, authorization and encryption would be done by publicly accessible PING servers. A key goal was to shift control of the record entirely to the patient and away from any institution or care provider.

As with Guardian Angel, the PCHR was not envisioned as a passive repository. An important component would be providing an 'app platform' as shown in Fig. 4.1 (courtesy of Dr. Ken Mandl), that would provide patients with useful tools to utilize their own data in ways that could assist them in managing their medical risks and any diseases they might have developed.

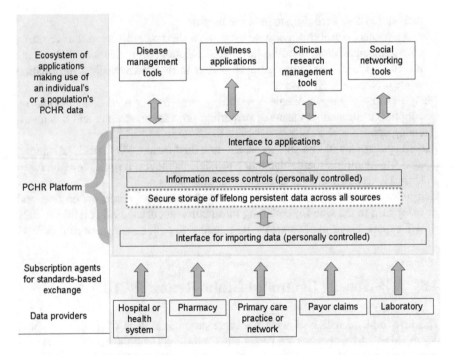

Fig. 4.1 Mandl and Kohane proposed a Platform Model of Personally Controlled Health Records (PCHRs) in a 2008 article in the *New England Journal of Medicine*. (http://www.nejm.org/doi/full/10.1056/NEJMsb0800220) The key points in this updated graphic are data aggregation from all of the patient's providers, persistent secure storage of the data, access control to the data, and an interface to an ecosystem of apps using data for a patient or for a population of patients. This schema is rapidly becoming the industry standard for both provider and patient apps through its later instantiation as SMART on FHIR. (Courtesy Dr. Ken Mandl)

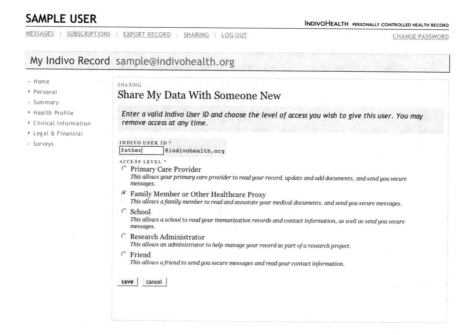

Fig. 4.2 This Indivo tool facilitates one of the goals of patient control of their health record by providing a facile mechanism for those patients to share their data with all of their providers, with family members, care givers or friends and for research purposes. (Courtesy Dr. Ken Mandl)

In an early demonstration of the potential for PCHRs in population health a group in Canada used PING as a prototype of a personally controlled record system for diabetes patients. The system allowed patients to register and create their PING records, transfer clinical data from various sources to that record and enter their daily step count as well as a portal to view and annotate the information in their record.

PING was succeeded by Indivo, a free, open source, standards-based tool that, for the first time, even though PING had a similar architecture, was called a patient-controlled health record (PCHR).[5] Figure 4.2 shows how the Indivo tool allows patients to designate their preferences for sharing of their data with all of their providers, with family members, care givers or friends and for research purposes. You might enjoy viewing the patient-oriented Indivo movie on the tool's site.[6]

Of course, the almost universal use of smartphones provides a near ideal, ubiquitous PHR platform. Since Indivo was developed by the same group that later created the SMART on FHIR app platform, it should also not be a surprise that, although the developers have ceased updating it, the last posted version of Indivo supports that platform allowing patients to utilize their health data stored in Indivo through the lens of whatever SMART on FHIR apps best meet their needs.[7]

---

[5] https://bmcmedinformdecismak.biomedcentral.com/articles/10.1186/1472-6947-7-25

[6] http://indivohealth.org/

[7] https://github.com/chb/indivo_server

For readers interested in more detail on their work and the evolution of the field in general, in 2016 Mandl and Kohane published an interesting perspective on the evolution of the personal health record.[8]

The original vision of a PHR was a patient-maintained record, but you can see from the options Indivo offers for data sharing that the scope of PHR use cases is expanding in our ever more connected world. Today, the Markle Foundation's "Connecting for Health" initiative defines a personal health record as "An electronic application through which individuals can access, manage and share their health information, and that of others for whom they are authorized, in a private, secure, and confidential environment."[9]

Despite technical progress, adoption of PHRs and other tools for patients remains an issue. A recent paper suggests that current adoption is around 20–30% of patients but forecasts that the majority of patients will use a PHR by 2020.[10]

There are several challenges to the adoption and successful use of PHRs:

- getting the data (interoperability)
- privacy concerns
- patient engagement
- usability

We will discuss each of these at a high level. For a more detailed discussion, I refer you to a recent paper that discusses PHRs and their challenges in some detail.[11]

Once again, interoperability is a key issue. Absent it, how does EHR data get into a patient's personal record? This becomes even more problematic if they receive care from several providers as patients with multiple chronic diseases typically do. Healthcare organizations have historically resisted meaningful data sharing out of concerns about privacy and security and for business reasons. The latter issue is called 'data blocking' and is the result of large health systems essentially wanting to lock-in their patients so they receive all care within the system by making it more convenient for the patient and by erecting barriers to providers outside the system obtaining needed data.

These same large healthcare systems often offer their patients the most common form of personal health record – a browser-based 'patient portal'. Enterprise software vendors often supply their clients with a portal and automatically populate them with data from their EHR and their other clinical systems. Figure 4.3 is an example of a modern patient portal provided by Marshfield Clinic Information Services (MCIS). This company is a spinout from the Marshfield Clinic, one of the most highly regarded health systems in the US and an advanced user of informatics. Note that it brings together a patient's health record, reminders of needed tests and procedures, messaging and financial and administrative services.

---

[8] http://www.nejm.org/doi/full/10.1056/NEJMp1512142#t=article

[9] Connecting for Health. The personal health working group final report. Markle Foundation; 2003 Jul 1.

[10] http://www.jmir.org/article/viewFile/jmir_v18i3e73/2

[11] https://www.ncbi.nlm.nih.gov/pmc/articles/PMC1447551/

Fig. 4.3 A modern patient portal provided by Marshfield Clinic Information Services (MCIS), a commercial spinout from the noted Marshfield Clinic. Data is automatically populated from the MCIS EHR and, as illustrated here, patients are provided with useful tools for both clinical and administrative tasks such as scheduling an appointment. (Courtesy MCIS)

One of the key goals of Meaningful Use is to assure that patients have electronic access to their records, no matter where they receive care. Meaningful Use has a term for this: View, Download & Transmit (VDT) and it is a key measure for providers. In Stage 2, more than 50% of all unique patients seen during each 90-day reporting period must have had timely online *access* to their health information. More than 5% *must have actually used* this capability. The second measure seems designed to incent providers to explain their VDT tool and encourage their patients to use it. Some research suggests that this provider engagement is effective. Hospitals and care providers even employ specialized patient engagement advocates/coordinators to serve as "a translator, or guide, with the abilities and skills necessary to assist patients with their health information."[12]

There are also Meaningful Use measures relating to sharing of health data at transitions of care. The Continuity of Care Document (CCD), an HL7 specified, XML formatted electronic patient summary, is one way to meet this requirement. As you will see in the HealthVault exercise that follows, it is also a way for a patient to obtain and work with all their EHR data from all their providers into a PHR.

The second challenge is patient concerns about the loss of privacy if their health data is stored 'in the cloud'. In a 2008 Markle Foundation survey of 1,580 adults, 77% reported privacy concerns related to misuse of personal data by marketers, 56% expressed concerns about misuse by employers and 53% had similar concerns about insurers.[13] The ACA provision outlawing the use of pre-existing conditions to

---

[12] http://bok.ahima.org/doc?oid=107741#.WgS3QmhSyUk

[13] http://www.connectingforhealth.org/resources/ResearchBrief-200806.pdf

deny health insurance may have alleviated these concerns to some degree with respect to health insurers although all of this is subject to change based on the shifting politics of healthcare.

However, data breaches are increasingly common, including in healthcare, so this issue is likely to remain a serious challenge. One particularly significant example was a breach by criminal hackers into the servers of Anthem, the largest for-profit managed health care company in the Blue Cross and Blue Shield Association. It involved the potential theft of over 78.8 million records containing personal information.[14] While Anthem says no medical information was compromised the information at risk contained names, birthdays, medical IDs, social security numbers, street addresses, e-mail addresses and employment information, including income data. This information could be used for identity theft or to file false medical claims.

Finally, there is the challenge to get patients to use these tools. Several studies have asked patients what functions they most want from a PHR. Unsurprisingly these usually include access to their past medical history including immunizations, lab tests and medications as well as access to their appointments and communication with their providers. Patients often want to access their family members' records. Age, lack of computer literacy, and unfamiliarity with medical terminology are often impediments to PHR use.

One interesting study summarized the results of five focus groups as indicating that patients wanted "(1) novel content that was relevant to their immediate and ongoing care, (2) a PHR they could trust for accuracy, security and privacy and (3) a highly functional PHR, facilitating care and communication with their clinician, and providing access to comprehensive personalized information shared with the clinician. Although practical usefulness was said to be essential, a major reason why participants said they trusted, used and sought relevance in the PHR was that it was offered to them by their personal clinician."[15]

4.4   PHR Exercise

If you have no personal experience with a PHR or have only used one or more portals from your healthcare provider(s) you may wish to follow along in this discussion using Microsoft's HealthVault PHR to gain a personal exposure to this technology.[16] HealthVault is free, public and does a good job of demonstrating some key PHR concepts, capabilities and challenges.

Your experience will be enhanced if you can obtain your patient record from your provider(s) as a CCD or as a CCR, its predecessor format. It is possible that your provider's portal offers these options. If not, ask. You can also upload your data

---

[14] https://www.wsj.com/articles/anthem-hacked-database-included-78-8-million-people-1424807364
[15] Kerns JW, Krist AH, Longo DR, et al. How patients want to engage with their personal health record: a qualitative study. BMJ Open 2013;3:e002931.
[16] https://www.healthvault.com/

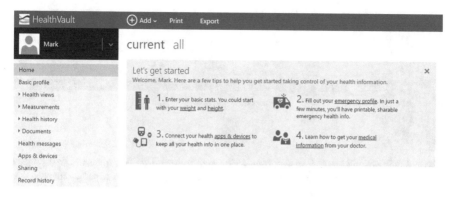

Fig. 4.4 HealthVault provides a friendly interface that walks non-technical users through getting started and offers upload options to lessen the significant and often problematic task of getting started by initializing the PHR with current health data. (Courtesy HealthVault)

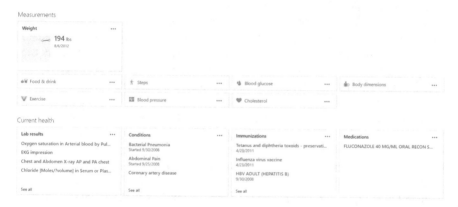

Fig. 4.5 HealthVault displays the key health data each user wants as tiles on a user customizable 'dashboard'. (Courtesy HealthVault)

in a spreadsheet or word processing format. We will show how to do this in one of the following paragraphs.

In Fig. 4.4 you can see that using HealthVault starts with recording or uploading the common health information patients say they want to keep track of. You can upload data from numerous personal devices and apps as well as from your providers if it's in one of the formats we just discussed.

As shown in Fig. 4.5, HealthVault displays the health record in tiles. Each user can decide which ones they want on their personal health 'dashboard'.

Earlier we discussed the possiblity of uploading your health data into HealthVault. Figure 4.6 focuses on the Documents menu choice where this is done. It expands when chosen to show support for Continuity of Care Documents (CCDs) along with the older CCR format and files from spreadsheets or word processing software. In addition to structured information that can be loaded into the PHR, users can also upload images. For more on that refer to the Health Information section of HealthVault's Help page.

Fig. 4.6 HealthVault can upload data from a variety of formats and that data would be displayed in the tiles and elsewhere. Users can even store medical images, scans of paper documents, etc. (Courtesy HealthVault)

Note that this make believe patient has six CCDs that could have come from six different providers caring for their various problems. This is an illustration of the potential of a PHR to be the place where all of a patient's data comes together.

The Meaningful Use Stage 2 objectives for care coordination and patient engagement required the latest HL7 document standard format (called Consolidated Clinical Document Architecture or C-CDA) so all stage 2 certified EHRs can produce C-CDA documents.[17] CCDs are a specific template within the C-CDA standard that is well suited for loading data into a personal health record.

Any software that supports the standard, as HealthVault does, can upload and should generally be able to correctly store most data from CCDs created by the patient's provider's EHR. HealthVault does this when patients push the Add Items button. If you have uploaded a CCD you can do this by highlighting it and then clicking on the ... link to bring up a menu. Importantly, most browsers will render XML documents in reasonably human readable form. In Fig. 4.7, you see the CCD highlighted in Fig. 4.6 after clicking the View Details link in the menu (Fig. 4.8).

Since patient's express concern about privacy they might well want to choose what information they wish to share. Spend a few minutes looking at the choices and you should quickly appreciate why understanding medical terminology is one of the challenges patients face in using a PHR. Now, spend some time exploring the various sharing options offered on this page, do you agree that computer literacy might also be a barrier to some patients using a PHR? Finally, do you think that there might be even easier, more usable ways for patients to express these preferences and choices?

If you own a device or use an app that can record health data, try to upload data from it into HealthVault. There is an extensive list of supported apps and devices so your current app or device may connect to HealthVault. Figure 4.9 gives an example of this done by the author using the Moves app on his Android smartphone to track activity data including time spent walking, step counts and other data, such as calories burned. There are two obvious advantages of this. Data can be displayed more clearly and for larger time periods on a larger device, and it can be correlated with

---

[17] https://www.healthit.gov/sites/default/files/c-cda_and_meaningfulusecertification.pdf

Title: Continuity of Care Document
Created On: August 26, 2014

Patient Information

Patient:	Josephine Vazquez		MRN:	527-67-0000
	5321 Timberlux Circle		Sex:	Female
	Atlanta		Race:	Black
	GA		Ethnicity:	
	30380		Language:	eng
	US		Next of Kin:	
	tel:+1-607-335-0471		Marital Status:	
Birthdate:	April 14, 1975			
Guardian:				

Care Team

| Name | Contact Information | Relation |
| Yoko Carroll | tel:1-740-597-5717 | Emergency Contact |

Discharge Information

Admission Date:	August 26, 2014
Discharge Date:	August 26, 2014
Discharge Disposition:	
Discharge Location:	Michael R Kletz MD

ADVANCE DIRECTIVES

Directive	Description	Custodian	Status
Resuscitation	Allow Resuscitation	Michael R Kletz MD	Supported By Durable Power of Attorney for Healthcare
IV Fluid and Support	Allow IV Fluid and Support	Michael R Kletz MD	Supported By Durable Power of Attorney for Healthcare
CPR	Do not allow CPR	Michael R Kletz MD	Verified With Family Only
Intubation	Do not allow Intubation	Michael R Kletz MD	Supported By Healthcare Will
Life Support	Do not allow Life Support	Michael R Kletz MD	Verified By Medical Record Only

ALLERGIES

Type	Substance	Reaction	Status
No known drug allergies			

FAMILY HISTORY

Relation	Diagnosis	Age At Onset
No family history data for this patient		

IMMUNIZATIONS

Vaccine	Date	Status
No vaccine data is available		

Fig. 4.7 HealthVault can display an uploaded CCD in a human readable form. This is one of the advantages of the XML standard. (Courtesy HealthVault)

other sources of data that are also available in HealthVault. Finally, this patient generated data could be shared with the patient's family or providers using HealthVault's data sharing tools.

4.5 Medlio's FHIR App

Might a FHIR enabled ecosystem of EHRs make creating and maintaining a PHR easier for patients? Several commercial vendors have presented FHIR-based app solutions based on this premise. They allow patients to aggregate data from all their providers. First, of course, patients must identify their providers to the app. Medlio is one of these new companies. Its app links to a directory containing 4.2 million provider profiles. To make searching easier, patients are defaulted to their current location but can search anywhere. As shown in Fig. 4.10, once a patient has selected

Invite someone to access Mark's information

✉ SEND INVITATION   ✕ CANCEL

Important: Some information stored in the records you manage may be highly sensitive. Before you grant access to a record, consider carefully which people should be allowed to see the information. Learn more about sharing records.

* Recipient's email address:

[   ]

* Retype email address:

[   ]

Passcode (optional)

[   ] (Minimum 4 characters)

If you create a passcode, the email recipient will need to enter it to accept this invitation. To protect your invitation, don't email the passcode. Use another method to tell it to the recipient. If you forget the passcode, you'll need to resend the invitation with a new one.

Sharing level
○ View Mark's information
● View and modify Mark's information
○ Act as a custodian of Mark's record (What can a record custodian do?)

Information Types
○ Share all types of information
● Share only the types of information selected below

┌─ * Information Types   select all | clear all ─────────────────────────────────────
│ ☑ Action Plan   ☑ Education - MyData file (preview)   ☑ Meal definition
│ ☑ Advance directive   ☑ Education - SIF student academic record (preview)   ☑ Medical annotation
│ ☑ Aerobic profile   ☑ Education document (preview)   ☑ Medical device
│ ☑ Allergic episode   ☑ Emotional state   ☑ Medical image study
│ ☑ Allergy   ☑ Encounter   ☑ Medical problem
│ ☑ Ambient temperature   ☑ Exercise   ☑ Medication
│ ☑ Application data reference   ☑ Exercise samples   ☑ Medication fill
│ ☑ Appointment   ☑ Explanation of benefits (EOB)   ☑ Menstruation
│ ☑ App-specific information   ☑ Family history   ☑ Message
│ ☑ Asthma inhaler   ☑ Family history condition   ☑ Microbiology lab test result
│ ☑ Asthma inhaler usage   ☑ Family history person   ☑ PAP session
│ ☑ Basic demographic information ☑ File   ☑ Password-protected package
│ ☑ Blood glucose   ☑ Food & drink   ☑ Peak flow

Fig. 4.8 HealthVault allows patients to select to share only specific subparts of their health record. This is called 'data segmentation' and providing a simple and clear way for patients with limited familiarity with medical terminology to specify it is a long-standing challenge. (Courtesy HealthVault)

a provider, they can choose what data and date range they want to load into their app. To actually obtain that data, they must first logon using their credentials for the chosen provider's patient portal.

One arm of the Medlio app seeks to create a meaningful medication list using FHIR Medication Order Resources from only the providers who wrote the prescription. However, every new refill for a prescription creates a new FHIR medication order resource that makes this a long and redundant list. The company's solution is to reconcile the orders using the RX Norm CUI number, a 6-digit numeric code that very specifically represents medications. We will discuss RX Norm later. They then group the orders based on this 6-digit number so the list shown in Fig. 4.11 contains each medication only once. Patients can see each prescription, the dates, and the prescribing physician(s).

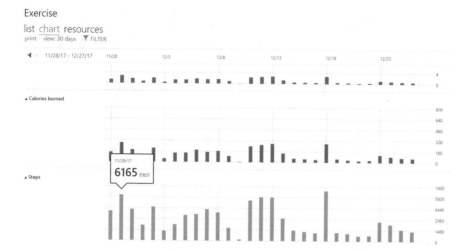

Fig. 4.9 HealthVault graphically displays data imported from the Moves app on the author's smartphone. Shown here are the daily step count and the calories burned based on the author's gender, age, height and weight. Both of these are calculated from data obtained from sensors built into the phone and might be different if the data came from another app on that same phone. For example, as you can see from the graphic, Moves calculated 6,165 steps on November 29th. The separate Google Fit app running on the same Android phone calculated 7,818 steps demonstrating one of the challenges of mHealth data we will discuss later on – its accuracy, particularly when it is calculated rather than directly measured.

4.6   CareEvolution's FHIR Based PHR

CareEvolution provides secure interoperability solutions for population health management, care coordination, and consumer engagement. The company offers a platform technology based on FHIR, other HL7 standards, IHE profiles (that, similarly to FHIR profiles that we will discuss later, seek to provide precise definitions of how standards can be implemented to meet specific clinical use cases) and secure messaging. The company says their goal is to enable organizations to obtain, standardize, and aggregate clinical and claims data into a single standards-based repository from which they can run web and mobile applications for clinical and operational analytics, quality measures, care management and to provide a provider portal and patient engagement tools to support quality improvement and transition to value-based care models. The company says their platform has been adopted by major health systems, integrated delivery networks, health plans, and research institutions that currently use it to manage data for over 130 million U.S. consumers.

CareEvolution leverages their own platform to offer a FHIR-based PHR solution. The company says that it provides secure interoperability solutions for population health management, care coordination, and consumer engagement. Its myFHR™ (Family Health Record) App runs on the web and on iOS devices and leverages FHIR to enable consumers to aggregate and manage their health data from

Fig. 4.10 Once they have
selected a provider,
patients using Medlio can
chose what data they wish
to download into their
PHR. (Courtesy Medlio)

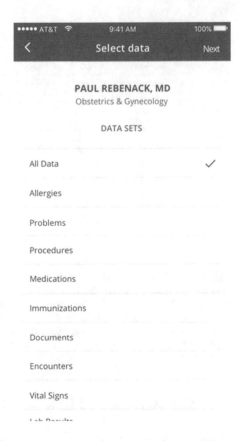

multiple hospital/physician office portals. Consumers can view and manage their medications, allergies, and lab results. The app also supports wearable sensors and connected devices such as glucometers, weight scales and blood pressure monitors. It provides notifications for upcoming appointments; overdue preventive screenings and immunizations; and drug-drug interactions.

4.7   Apple's FHIR-Based Health App

As this book was nearing completion Apple announced an "updated Health Records section within the Health app brings together hospitals, clinics and the existing Health app to make it easy for consumers to see their available medical data from multiple providers whenever they choose." The Apple announcement listed 12 well known healthcare organizations that were participating in the beta test program for the new software. The capability is implemented in the March, 2018 11.3 version of iOS using FHIR.[18]

---

[18] https://www.apple.com/newsroom/2018/03/ios-11-3-is-available-today/

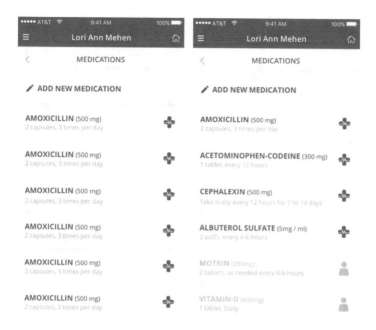

Fig. 4.11 The Medlio app uses RxNorm, one of the key health data standards, to convert the redundant list of FHIR resources on the left, where each medication dispensing is separately listed, to the more useful list on the right where each medication is listed only once. The use of RxNorm allows the app to recognize different brands of a medication as the same thing. (Courtesy Medlio)

In describing the new software Apple said "consumers will have medical information from various institutions organized into one view covering allergies, conditions, immunizations, lab results, medications, procedures and vitals, and will receive notifications when their data is updated." The company went on to say that "Health Records data is encrypted and protected with the user's iPhone passcode."

Screen shots from the Apple Health app are shown in Fig. 4.12. The shot on the left illustrates that patients see an integrated view of their medical records from multiple providers. The shot on the right shows how the data is organized for the user in what the company describes as "a clear, easy to understand timeline view."[19]

This development seems likely to accelerate the trend toward patient (and consumer) control of their own health data that, in turn, will likely lead to faster adoption of FHIR by health systems.

Currently Apple is supporting the key clinical FHIR Resources: allergies, conditions, immunizations, labs, medications, and procedures as well as the FHIR vital signs profile. All of these are assumed to be as specified by the 2015 Common Clinical Data Set (CCDS) as defined by ONC.[20] Because the CCDS specifies the

---

[19] https://www.apple.com/newsroom/2018/01/apple-announces-effortless-solution-bringing-health-records-to-iPhone/

[20] https://www.healthit.gov/sites/default/files/2015Ed_CCG_CCDS.pdf

Fig. 4.12 Screenshots from the Apple Health app illustrate that data is brought together from multiple providers (left) and can be viewed by the user in what the company claims is a clear to understand timeline view (right). (Courtesy of Apple)

data required certification 2015 Edition certification, it is of obvious interest to EHR vendors that must support it so it is included in the Argonaut Profiles. We will discuss FHIR profiles later on but this is a logical thing for Apple to have done since the group that created Argonaut was focused on this patient data aggregation use case to meet the requirements of Meaningful Use for patients to access their health records. As a result, Argonaut provides a significant degree of consistency to the data and FHIR formats that Apple receives. This implies that Apple need do nothing more to normalize the data. This could change should Apple expand the resources it wishes to support beyond those specified in Argonaut (for example, genomic data).[21]

Shortly after the Apple announcement Isaac Kohane, MD, PhD, Director of the Boston Children's Hospital Informatics Program that developed SMART on FHIR, wrote an article exploring the potential impact of the Apple enouncement. He began by saying that the announcement "is less than meets the eye because:"

> The data enabled is a mere trickle compared to the torrent of health care data that we all generate during our medical visits.
> It's also only a uni-directional data flow from the health care institutions – the hospitals, the medical practices – to your personal health record. Data will not flow in the other direction. You cannot update an incorrect observation in your health record, nor can you add missing facts or missing medications to your "official" health record. At least, not yet.

---

[21] http://argonautwiki.hl7.org/index.php?title=Implementation_Guide

It's also not a magic switch that will allow everyone access to their health care records. It requires that hospitals agree to work with Apple to provide this data at a reasonably timely interval or on demand. Currently, only a small number of hospitals have agreed to do so.

He then goes on to say that the announcement could nevertheless have major impact "Because it represents the first time a mass consumer platform that is in the hands of tens of millions of consumers daily and for hours on end – the iOS operating system – will get officially sanctioned health care observations from the formal institutional health care system. This development seems likely to accelerate the trend toward patient (and consumer) control of their own health data that, in turn, will likely lead to faster adoption of FHIR by health systems."[22]

To explore that in more detail and learn where it might lead I talked with Peter Celano, MBA, Director of Consumer Health Initiatives, at the MedStar Institute for Innovation, part of MedStar Health, an integrated delivery system in Washington DC and surrounding areas. He sees a future in which patients who are in control of their health record will seek care from the high convenience/low friction sources. Today, as we will discuss later in this chapter, such services are already available but they generally do not have access to each patient's actual medical record. Once patients are in control of that record, he feels they will inevitably inject their digital data into synchronous and asynchronous care platforms when pursuing screening, diagnosis and therapy. And he believes these services will increasingly use analytics to tailor personalized decisions about each patient's care. Physicians will likely be involved on an "over-read basis," but their role might be partially limited to those cases where the analytic system cannot be sure of what to offer based on the available information. Finally, Celano believes that for many mild conditions, a synchronous virtual visit (2-way audio/2-way video) might indeed be overkill, and the patient will simply be presented a short form and/or access to a chatbot, to at least start the encounter online.

To explore another interesting approach to patient access to their data, I talked with Alistair Erskine, MD Chief Informatics Officer, at the renowned Geisinger Health System in central Pennsylvania. Geisinger was one of the very first health care organizations to develop FHIR apps (we discussed their arthritis management app earlier) and has long been an exemplar of value-based healthcare delivery.

He told me that for a few months before the Apple's announcement Geisinger has allowed patients to access their health record as FHIR resources using a FHIR app *of the patient's choice* once they authenticate themselves to Geisinger's MyGeisinger patient portal (supplied by Epic) using their portal credentials via the Geisinger OAuth2 authorization server.[23]

Geisinger, of course, doesn't promise that the patient selected apps will prove to be of value or even function as advertised. To date patients have chosen four FHIR apps to connect: the myFHIR PHR app we discussed in the previous section, the Care Passport PHR app[24], the Apple Health app[25], and the 1upHealth platform for

---

[22] http://www.wbur.org/commonhealth/2018/01/26/apple-health-care-data

[23] https://mygeisinger.geisinger.org/mygeisinger/login.cfm

[24] https://carepassport.com/

[25] https://www.apple.com/ios/health/

exchange of electronic health data between patients, providers, and app developers).[26] To date only a few patients are participating and Geisinger won't widely promote the service until iOS 11.3 is fully released (at present it is in a beta test program).

Patients who do participate have the interesting ability to merge their Geisinger health record data with their own mHealth data either acquired by their phone or communicated to it from a device of the patient's choosing. Dr. Erskine points out that, ironically, patients who obtain all their care from one organization, as is typical for Geisinger patients, may not benefit as much from the access to their own data on a mobile device since they can already see it all using the Geisinger portal or mobile apps. Those patients whose care is provided by several health care systems might see greater benefit because they would, perhaps for the first time, have a complete view of their health record.

Of course, as you now know, personal health records that bring together all the data from the providers who may care for a patient are not a new idea. To date, it has not achieved a high degree of acceptance. Yet another example of this is Google Health a personal health record service introduced in 2008 and discontinued in 2011. Will Apple's announcement change that?

Dr. Ida Sim, a co-director of biomedical informatics at the University of California, San Francisco Clinical and Translational Sciences Institute believes that it will catch on this time because, unlike with Google Health, patients no longer have to enter their own data and, unlike before, virtually all providers have digital record systems (that wasn't true in 2008 when Google launched its service) and, as a result of the Twenty First Century Cures Act of 2016, there is a data standard for personal health records (that is likely to be formalized as FHIR once a normative version is released by HL7).

However, she feels that adoption and use of the data still remains a challenge. In an interview with NPR she said "We'll probably see huge numbers of people getting their initial Health Records populated. The issue is, then what? The value will come from third party apps that use Health Records to provide meaningful value to patients, and until this value is demonstrated, I think Health Records uptake will be large but retention and continued engagement of patients will be challenging."[27]

After the book was in production Apple announced support for the SMART on FHIR app platform suggesting that patient facing FHIR apps running on an iPhone would be able to access the medical record data and presumably any data recorded on the phone.

4.8   The CMS Blue Button 2.0 API

We discussed Geisinger's use of its patient portal to give patients access to their data using any app of their choosing. We also discussed that the data from such a portal would usually be limited to a single health system. We also looked at how a PHR or

---

[26] https://1up.health/

[27] https://www.npr.org/sections/health-shots/2018/02/14/585715952/medical-records-may-finally-be-coming-to-your-apple-smartphone

a FHIR app can make this task more manageable across providers. Could it be even easier for a patient to see all their data no matter where they receive care?

Since they pay for most of it, each patient's health insurance company should have a comprehensive view of their care, no matter where it takes place. This healthcare claims data has limitations. For example, while claims would show when a lab test was performed, they would not have the result. However, it also has clear advantages since it may often be the most comprehensive longitudinal view of a patient's care that is practically available.

The Medicare program is operated by the Centers for Medicare and Medicaid Services (CMS) on behalf of over 44 million beneficiaries over age 65. It is by far the largest health insurance program in the US. CMS has recognized the potential of its claims data if it is wrapped in innovative patient-facing apps. Its Blue Button 2.0 API initiative is based on FHIR and is designed to do just that by providing data access that will allow patients to access the apps, services, and research programs they find of value and trust.

The Blue Button 2.0 API was unveiled at HIMSS 2018. It was designed to

- Allow an app developer to register a beneficiary-facing application
- Enable a beneficiary to grant an app access to 3 years of their Medicare claims information
- Use the FHIR standard as a format for beneficiary information.

To facilitate app development the API provides for retrieving:

- A User Profile with appropriate access rights
- All Explanation of Benefit records for an Individual Beneficiary
- All Patient Records for an Individual Beneficiary
- Coverage information for an Individual Beneficiary[28, 29]

4.9   OpenNotes

As an alternative to any of these approaches for giving patients *the data* from their EHR records, why not give them direct access to those complete records? Budd N Shenkin, MD and David C Warner, PhD first proposed this 'radical' idea in the 1973 *New England Journal of Medicine*.[30] In 2010 Beth Israel Deaconess Medical Center in Boston, Geisinger Health System in rural Pennsylvania, and Seattle's Harborview Medical Center launched an exploratory study of the idea funded by the Robert Wood Johnson Foundation. Primary care doctors invited their patients to read their notes via secure online portals. The prestigious journal *Annals of Internal Medicine* published the results in 2012.[31]

---

[28] https://dev.bluebutton.cms.fhirservice.net/devdocs/

[29] https://hhsidealab.github.io/bluebutton-developer-help/

[30] http://www.nejm.org/doi/pdf/10.1056/NEJM197309272891311

[31] http://annals.org/aim/fullarticle/1363511/inviting-patients-read-doctors-notes-quasi-experimental-study-look-ahead

According to this effort, now called OpenNotes, the big takeaways were:

- Doctors report little change in workload
- Patients overwhelmingly approve of note sharing
- Few patients are worried or confused by their notes

Today five million patients have access to OpenNotes. You can visit the OpenNotes site for more information and you can also watch two brief videos posted by OpenNotes.[32] In the first video patients report that reading notes engages them and helps them feel more in control of their health and health care.[33] The second video provides evidence that patients can provide quality control for their own notes.[34]

4.10   PatientsLikeMe®

The company was founded in 2004 by three MIT engineers: Benjamin and James Heywood (brothers) and their longtime friend, Jeff Cole. Five years earlier, the Heywoods' brother, Stephen, had been diagnosed with amyotrophic lateral sclerosis (ALS, also known as Lou Gehrig's disease) at the age of 29. The family soon began searching the world for ideas that would extend and improve Stephen's life. Based on this experience they saw the need for an environment for sharing and collecting data, in their case focused on innovative treatments for an incurable disease. To accomplish this, they built a unique social networking system whose primary mission was to get people connected so they could share their experience to improve outcomes.

The site now has well over 2,800 conditions, but prior to April 2011 there were only 20. It is free to patients and accepts no advertising, but it is a for-profit business. The objective is to gather data from patients about their illness experience and make it available in aggregated form to other patients and to organizations that are interested in particular populations of patients. Examples would be pharmaceutical companies or companies with other early stage medical products that want to learn from patients that have the condition they seek to treat. For example, a pharmaceutical company might partner with the site to create a community for engaging organ transplant recipients where it can conduct research and learn from the patients, while at the same time considering their aggregated data.

The company also seeks to work to help design clinical trials that have patients' interests in mind. It says a good recent example of this is their involvement with Dr. Rick Bedlack of Duke University in a clinical trial of Lunasin, a 43-amino acid peptide originally isolated, purified and sequenced from soybean seed in 1987. It has been studied as possible drug for the past 20 years and more recently it has been suggested that it might reduce the progressive muscular weakening found in

---

[32] https://www.opennotes.org/

[33] https://www.youtube.com/watch?v=NnSF_Itu5cY&feature=youtu.be

[34] https://www.youtube.com/watch?v=WKl8qJuHJfU&feature=youtu.be

ALS. In a brief video, the company's Vice President of Innovation, Paul Wicks, discussed how patient communities can become involved in research.[35] The Lunasin study was the subject of a segment on the National Public Radio (NPR) program "All Things Considered."[36]

To create a clinically relevant research platform, PatientsLikeMe uses structured surveys to collect patient-reported data. Novel treatment, symptom, and condition data enter the "User Voice dashboard," where it is reviewed and curated to assure data integrity. The company receives around 75 "user voice" entries per day.

Some may already be in the system. For example, there could be a spelling difference, or the patient could have entered two concepts together, such as "pain and depression." The spelling error would be recognized to avoid duplicate concepts and the combined concepts would be split so the patient can monitor each separately and each can be aggregated for research purposes. All clinical data is coded in the background using standardized terminologies. Symptoms and side effects are coded into SNOMED-CT and the Medical Dictionary for Regulatory Activities (MedDRA), a medical terminology used to classify adverse events associated with the use of biopharmaceuticals and other medical products. Diagnoses are coded into ICD-10.

Despite this high degree of coding, as much as possible of the "patient voice" is maintained. PatientsLikeMe points out that the patients self-manage most of their care. As shown in Fig. 4.13, PatientsLikeMe allows patients to report symptoms and how they attempt to treat them. Their collective information, as shown in this example, can be of value to patients in coping with a complex and highly variable disease such as Systemic Lupus Erythematosus. It can also be of value to researchers interested in the natural course of a disease in a large population of patients affected by that disease.

4.11 Telecare

As should now be clear, chronic diseases drive most US healthcare costs. Once diagnosed, chronic diseases are usually not curable. After their diagnosis, patient behavior and compliance with the prescribed treatments is critical to success in managing them. As a result, there has long been interest in the potential for technology in the home to create a more continuous and coordinated approach to the care of these patients.

This is another old health informatics idea only now coming to fruition. Figure 4.14 is a mid-1990's photo of Steve Kaufman, one of the early innovators in this space, with his Home Assisted Nursing Care (HANC) 'robot'. HANC offered a wide variety of voice-controlled nursing services to patients at home. Among other things, HANC dispensed medications at the proper time and took physiologic

---

[35] https://www.youtube.com/watch?v=QxjJO_Uf860

[36] https://www.npr.org/sections/health-shots/2016/10/25/499328778/simplified-study-aims-to-quickly-test-a-long-shot-als-treatment

Common symptoms reported by people with systemic lupus erythematosus

Common symptoms	How bad it is	What people are taking for it
Stress		Listen to Music, Prayer/meditation, Relaxation Techniques
Fatigue		Rest, Amphetamine-Dextroamphetamine, Modafinil
Muscle pain		Tramadol, Methocarbamol, Carisoprodol
Joint pain		Hydroxychloroquine, Prednisone, Meloxicam
Headaches		Ibuprofen, Acetaminophen-aspirin-caffeine, Topiramate
Pain		Tramadol, Hydrocodone-Acetaminophen, Gabapentin
Joint swelling		Prednisone, Acetaminophen (Paracetamol), Celecoxib
Raynaud's phenomenon		Amlodipine, Nifedipine, Nitroglycerin topical
Skin sensitivity to sun (photosensitivity)		Avoid sunlight, Hydroxychloroquine, Quinacrine
Chest pain		Nitroglycerin, Electrocardiogram (EKG), Metoprolol
Anxious mood		Alprazolam, Clonazepam, Lorazepam
Depressed mood		Duloxetine, Sertraline, Bupropion

Reports may be affected by other conditions and/or medication side effects. We ask about general   Last updated: December 26, 2017
symptoms (anxious mood, depressed mood, fatigue, pain, and stress) regardless of condition.

Fig. 4.13 PatientsLikeMe patients can report symptoms and how they attempt to treat them. Their collective information, as shown here, can be of value in better understanding a complex and highly variable disease such as Systemic Lupus Erythematosus. (Courtesy PatientsLikeMe)

Fig. 4.14 Steve Kaufman invented HANC, a sophisticated but large and expensive in-home 'robotic' device that could advise patients, dispense medications and take physiologic measurements using built-in devices. (Courtesy Steve Kaufman)

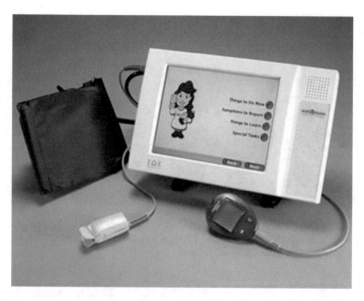

Fig. 4.15 The author was involved in the development of a system that provided similar functionality to HANC but took advantage of the miniaturization of components to do that in a cigar box sized device.

measurements, such as blood pressure, with its patient operated devices using the technology of the day.

During this same period, the author was one of the founders of a company that developed an electronic record system for the home care nurses who took care of patients after they left the hospital. The company could see the benefit of knowing more about the patients between nursing visits and of reinforcing the instructions the nurses provided. The device we developed was roughly cigar box sized and featured a color touch screen, as shown in Fig. 4.15. Like HANC, it spoke to the patients and had integrated physiologic measurement devices. Given the advances in technology, it was quite a bit smaller and less expensive than HANC but, nevertheless, adoption was limited because the payment and care models had not yet advanced in ways that made in-home technology economically viable for providers.

Of course, as you would expect, software for home care is now available on smartphones. Despite the huge reductions in size and cost, these devices are considerably more powerful and facilitate features that would have been unimaginable, or at least impractical, given the computer resources available for in-home use when these early systems were developed.

The example of telehomecare on a smartphone in Fig. 4.16 is from AlayaCare, a Canadian company.[37] Patients record some of the data. This example question would

---

[37] https://www.alayacare.com/

Fig. 4.16 Much of the functionality of the early telehomecare devices can now be provided on a smartphone for a fraction of the cost. Here a patient records information using an attractive and easy to understand format. (Courtesy AlayaCare)

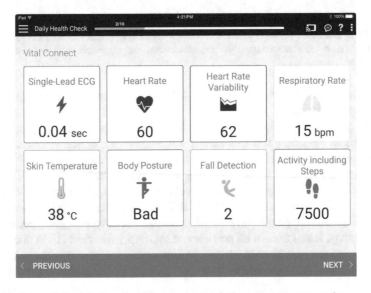

Fig. 4.17 External physiologic and activity monitoring devices are now commonplace, very compact and relatively inexpensive. Some measurements, such as activity and sleep, can be taken by sensors embedded in the phone. Others can be taken externally and synced to the phone via Bluetooth. (Courtesy AlayaCare)

be part of something similar to a Likert survey that collects the client's point of view on a daily basis.

The collection and display of vital signs, physiologic and activity data is illustrated in Fig. 4.17. It is no longer necessary to provide integrated physiologic devices since they are now consumer products and most have wireless Bluetooth syncing capability. The company says it can synchronize data from over 200 devices. It brings this data together for quick review by the nurse. This data informs an algorithm used in the dashboard we will discuss next.

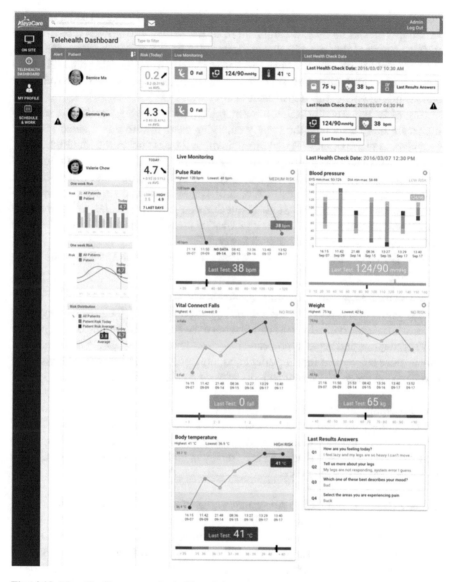

Fig. 4.18 The AlayCare system's dashboard features patient-specific upper and lower bound thresholds that trigger workflows for various care team members. If a threshold boundary is detected, a risk score is calculated using rules developed with machine learning. (Courtesy AlayaCare)

Finally, Fig. 4.18 illustrates a dashboard that provides nurses with a quick overview of their patients. It provides a graphical representation of the incoming data from various devices. Importantly, since a value might be normal for one patient but abnormal for another, each measure has patient-specific upper and lower bound thresholds that trigger workflows for various care team members. If a threshold boundary is detected a risk score is calculated using rules developed using machine learning. The risk score is a combination of the patient's diagnoses (coded in the ICD 10 standard that we will discuss later), demographic information, vital signs data and the Likert survey. AlayaCare has trained their model using data sets from their remote monitoring partners that, critically, include events such as hospital re-admissions or emergency department visits.

Arguably nothing illustrates the emerging and expanding landscape of tele-homecare more fully than the rapidly emerging trend toward direct video visits by providers to their patients. The platform is often the now ubiquitous smartphone.

Dr. Robert Pearl, former CEO of The Permanente Medical Group, the nation's largest medical group with over 9,000 physicians and 35,000 nurses who staff the Kaiser HMO plans, and former President of The Mid-Atlantic Permanente Medical Group, wrote an interesting article about this new trend for *NEJM Catalyst* a new online publication intended to share innovative ideas and practical applications for enhancing the value of health care delivery.[38]

In it Dr. Pearl discusses several use cases for the technology at Kaiser under his leadership:

Tele-Dermatology   An interesting application of the technology since using a high quality photo of the skin condition made by the patient using their own smartphone camera "a remote dermatologist can diagnose the condition from the digital picture 70% of the time and begin appropriate treatment immediately." Dr. Pearl argues that this is more convenient for the patient and more efficient for the dermatologist who can focus on the other 30% of patients who *do* need to be seen in the office.

On-Site Specialty Consultation   Dr. Pearl says that at Kaiser a PCP can now use the technology for their patients who require a specialty consultation from any of Kaiser's more than 40 medical and surgical specialties. He argues that this application reduces the time to diagnosis and treatment and helps insure that the PCP and specialist coordinate their care.

Urgent-Care Virtual Visits   According to the CDC in 2017 there are 141 million emergency department visits annually in the US. Only around 8% of these result in a hospital admission, suggesting that many might not require urgent care. In fact, studies have consistently shown that this is the case.[39, 40] At Kaiser, patients who feel they have an urgent problem can be connected to a physician by phone, and 80% of those problems can be managed without a visit. 60% of the rest can be managed using a video visit. Over 90% of patients report being satisfied with receiving care this way.

Routine, Follow-Up, and Specialist Visits   All Kaiser patients are offered a video visit as an alternative to a physical one "for a variety of routine and urgent medical

---

[38] https://catalyst.nejm.org/engaging-physicians-in-telehealth/

[39] https://www.ncbi.nlm.nih.gov/pmc/articles/PMC3876304/

[40] https://www.ncbi.nlm.nih.gov/pmc/articles/PMC3412873/

problems, specialty consultation, and surgical follow-up." A third of patients voluntarily chose this option.

Finally, Dr. Pearls discusses the impact of this technology on the cost of care delivery. His conclusion about this may surprise many readers:

> That's not easy to quantify. On the one hand, the required physician and non-physician staff time is similar to that for an in-person visit. On the other hand, costs related to capital, transportation, parking, and time away from work are much lower. Likewise, although telehealth's convenience increases the overall number of visits, its speed also reduces delays in diagnosis and treatment – and helps to avoid longer-term complications and hospitalizations. Overall, telehealth has similar costs to that of traditionally delivered care, but the improvements in quality and service are major.

Today the idea of virtual smartphone-based physician visits is a commercial reality. Moreover, they don't just originate with health systems. Numerous for-profit companies have emerged offering smartphone access to physicians to patients who are able and willing to pay.[41] This service is even being offered directly to their policy holders as an alternative care model by commercial health insurance companies.[42] Perhaps of even greater significance is Medicare's decision to allow eligible practitioners to bill (using the CPT code 99091 for Miscellaneous Medicine Services) to receive separate reimbursement "for time spent on collection and interpretation of health data that is generated by a patient remotely, digitally stored and transmitted to the provider, at a minimum of 30 min of time." This may well be a first step toward recognizing remote patient monitoring services for separate payment. Historically the rest of the healthcare payment system follows CMS's lead so telecare seems destined to become a widely accepted and utilized treatment modality.

4.12   Recap

Providing patients with facile access to their own health data is at the core of many recent innovations in health informatics. Since the earliest days of the personal health record various approaches have been explored to help engage patients in their own care, to support the public to stay healthy, and to provide providers with a more complete and continuous view of their patients' longitudinal records. The various examples discussed in this chapter illustrate how the adoption of open, standards-based healthcare technologies is attracting innovators (and capital to support them) for the development of sophisticated and non-traditional approaches to some of healthcare's biggest and longest-standing challenges.

---

[41] http://www.mobihealthnews.com/34027/12-virtual-visit-services-that-connect-patients-at-home-to-doctors-or-nurses

[42] https://www11.anthem.com/inthenews/lho/

Chapter 5
Health Information Exchange

5.1 Introduction

In the previous chapter we discussed how health information technology is being used to help patients better manage their own care. In this chapter, we will focus on main impediments to interoperability and provide examples of how those impediments have been overcome, at least in part.

We will begin with a discussion of three levels of interoperability and provide an example of how HL7's Clinical Information Modeling Initiative (CIMI) attempts to address semantic interoperability, which is the most complex and, hence, difficult to achieve level. We'll also examine how these semantic interoperability concepts can be implemented in the real world through the use of tools like Applicadia, which uses voice recognition and automated encoding to turn physicians notes into structured data that can be analyzed, aggregated, and compared in a meaningful way. These tools demonstrate the potential viability of achieving semantic interoperability through machine learning and other technologies.

The emerging health informatics technologies we discuss in this book offer great promise. The opportunities for analytics create ever increasing demand for data. Unfortunately, significant impediments to interoperability still exist. Many of the policies and regulations that restrict the exchange of data, such as HIPPA, privacy and security rules, and trust agreements were designed in the pre-Internet era and they cannot be overcome through purely technologic solutions.

All of these factors are provided as background information before shifting to the main focus of this chapter: health information exchange, which is a specialized network for the secure sharing of patient data. We'll trace the history of HIE back to the early days of HL7 V2 messaging and demonstrate various HIE approaches that are currently in place or underway. We will end with a discussion of future directions where HIEs seem to be headings.

© Springer International Publishing AG, part of Springer Nature 2018
M. L. Braunstein, *Health Informatics on FHIR: How HL7's New API is Transforming Healthcare*, https://doi.org/10.1007/978-3-319-93414-3_5

5.2   The Interoperability Challenge

At the outset of the book we discussed the informatics challenges to achieving a Learning Healthcare System as adoption, interoperability and analytics. Adoption has largely occurred so the current pressing challenge is interoperability, the ability of systems to share information. These two challenges are connected. Widespread adoption has increased the focus on achieving interoperability in part because providers now using EHRs are frustrated that they cannot easily share data. The US Congress is concerned about achieving the expected benefits from the billions of dollars that the HITECH program invested in adoption without workable interoperability. Interestingly, as we will discuss later, widespread EHR adoption may provide training datasets for new, more automated approaches to interoperability. The announcement by CMS administrator, Seema Verma, at HIMSS 2018 is indicative of the degree to which interoperability has become US Federal government policy.

We have discussed EHRs and PHRs, key clinical systems that depend on interoperability to obtain and share data. Another key beneficiary from more open access to data is laboratory information systems that provide critical test results that can save lives *if* they get to the right place in a timely manner. Another are pharmacy information systems. They can spot inappropriate or even life threatening medication orders, but must obtain a complete medication list and have a way to warn providers and even patients when they do identify a potential problem. Of course, the value of mHealth and the data it can collect is tied to how easy it is for that data to be accessed by the systems providers and patients use.

The multiplicity of commercial EHRs and the desire of health systems to control and use patient data strategically are some of the main interoperability impediments. A third issue is the complexity of interoperability itself. To explore that, we will consider three interoperability paradigms or layers in order of increasing sophistication and, hence, from easiest to most difficult to achieve. Don't be concerned if the distinctions aren't clear to you now, later we will explore current standards and technologies to achieve each of these levels of interoperability.

Transport Interoperability The simplest and easiest form of interoperability is simple transport of data among systems. As with any communications there must be a standard for packaging and transporting the information that is well understood and implemented by both sending and receiving systems. Importantly at this level the transport mechanism has no knowledge or understanding of the data or of what it means. You use transport mechanisms like this on a daily basis. These include phone calls, faxes, and, most commonly, email.

Structured Interoperability In this next level data is packaged in a well-defined manner such that the identity of each element is understood by all systems based on where it is located in that structure. This allows data to be parsed when received and stored in the appropriate locations within the receiving system. A standard for this must identify the allowable values of each data element to ensure that the sending and receiving systems are speaking a sufficiently similar language. Simple transport interoperability can convey information that has a known and well accepted structure but it might also be sending a fax or scan image with no structure at all.

Semantic Interoperability This is the highest level of interoperability and it requires that the data is sufficiently standardized that its meaning will be clear to both the sending and receiving systems. A good example of what that implies would be that clinical data collected in EHR A could be reliably used by a clinical decision support module in EHR B once it has received that data.

At the present time transport and structured interoperability are widely used. FHIR is a structured interoperability standard. What about semantic interoperability which many would describe as the 'holy grail'?

5.3   Clinical Information Modeling Initiative: CIMI

In 2013 Grahame Grieve, the leader of the global HL7 FHIR effort, posted in reference to semantic interoperability: "let's not pursue the holy grail right now."[1] At present FHIR does not attempt to address semantic interoperability, in large part because of the *additional* complexity that would introduce to an already complex effort. I will give you a sense of that complexity here, but a complete treatment of the subject is well beyond the scope of this book. There are many strongly held points of view about the feasibility of semantic interoperability. There are also strong advocates for various approaches to a semantic interoperability standard. We will now discuss one of the latest approaches.

The Clinical Information Modeling Initiative (CIMI) is another HL7 working group conceived by the same 'fresh look' taskforce that birthed FHIR. Its mission is to improve the interoperability of healthcare systems through shared, implementable clinical information models. The ultimate goal is semantic interoperability.

It is important to understand that data exchange and the use of coding standards *alone* is insufficient to achieve semantic interoperability. As we will see, there can be many codes for a lab test or a clinical diagnosis. In the case of a lab test, codes could specifically indicate which of several available methods the lab used and this could affect the interpretation of the result. Other codes might indicate what kind of test produced the result. For a diagnosis, there can be many levels of coding detail and specificity. There can also be structural differences in coding systems. Providers can differ as to how and when they use these codes. Finally, as we will now discuss, multiple EHRs can address the representation of the same clinical concept differently.

Figure 5.1 presents a simple illustration of the variability of representation and of the semantic interoperability problem CIMI is trying to solve. Three EHR systems represent the concept of 'suspected lung cancer' differently. In the General Practice EHR on the left, it is a compound concept created by combining three structured

---

[1] http://www.healthintersections.com.au/?p=1407

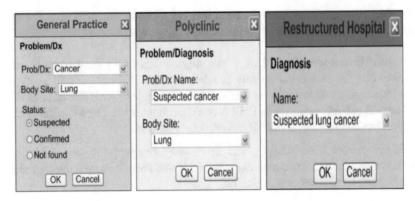

Fig. 5.1 Three EMR systems represent Suspected Lung Cancer in semantically different ways. The General Practice EMR represents it as three structured fields – Problem/Diagnosis, Body Site and Status. For the purpose of this discussion that is the preferred representation. The Polyclinic EMR represents it as two structured fields – one for the Problem/Diagnosis and one for its Body Site. The Restructured Hospital EMR represents it as a presumably free text Diagnosis. (Figs. 5.1, 5.2, 5.3 and 5.4 courtesy Stanley M Huff, MD, Chief Medical Informatics Officer at Intermountain Health)

Fig. 5.2 CIMI uses shapes and colors to represent the different ways our three hypothetical EMRs represent the same clinical concept. (Courtesy Dr. Stanley M Huff)

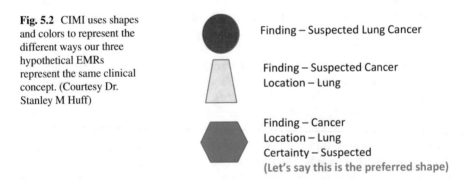

data elements. In the Restructured Hospital EHR on the right, it is a single text concept. The Polyclinic EHR in the middle takes an in between approach by adding Body Site to further specify the Diagnosis/Problem field. Technically there is not a right and wrong way to do this, but three different ways used in parallel make it harder to aggregate the data meaningfully for analysis or to share it to coordinate care among these three providers.

In Fig. 5.2 a CIMI graphic uses shapes and colors to represent the different ways the three EHR systems represent this single clinical concept. For the purposes of illustration, we are assuming the blue hexagon (the approach used by the General Practice EHR) is the preferred shape and a hypothetical translation service might put all the instances of this clinical finding into that shape. This is essentially a standard structure in which terms always go in the same place, something we will see in the HL7 messages we will discuss later on. FHIR Resources also provide a standard structure specifying where data elements are located. In that context, semantic

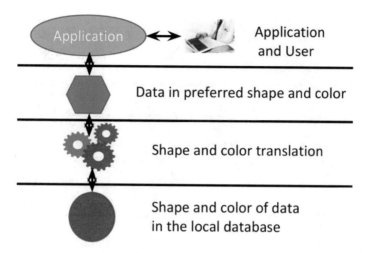

Fig. 5.3 Data is transformed from the red and yellow 'shapes' into the blue shape meaning that the fields used to describe aspects of the problem/diagnosis are now the same across the EMRs but the terms placed in those fields can still be different. (Courtesy Dr. Stanley M Huff)

interoperability specifies what formats and values to use in specific clinical scenarios, often called use cases.

In Fig. 5.3 middle ware transforms the shape and color of all the data into the preferred 'blue' representation. However, this doesn't achieve true semantic interoperability because, although the data is now in the preferred format, the terms used can still be different.

In Fig. 5.4, the most advanced CIMI proposed solution, data not only goes into a consistent location but the terms are also translated into an agreed upon standard as specified in a CIMI model. This translation is to some of the data standards we will describe in a later chapter. Now, as also illustrated in Fig. 5.4, a hypothetical FHIR app or other clinical tool gets the same structure *and terms* even though the source systems use different representations for both.

The goal of CIMI is to create an open shared repository of models to drive semantic interoperability. The details of modeling clinical data are far too complex for this book but Fig. 5.5 (from Dr. Linda Bird who served as chair of the modelling taskforce for CIMI) provides a simple illustration of what a model needs to achieve using the same 'suspected lung cancer' problem. The entries under our three EMRs illustrate their different representations of just three of the seven fields the model requires for a problem/diagnosis. Those entries and the others all need to be consistent.

This figure only hints at the complexity of actually implementing the full CIMI concept. Semantic interoperability is very difficult to achieve because of the complexity of healthcare data, the many ways in which it is represented and the many contexts in which it is used. We will discuss FHIR profiles later but they essentially constrain the standard for a particular use case. Thus, they could specify the 'preferred shape'. CIMI anticipates that its models will connect to and inform or constrain FHIR through model-derived FHIR profiles.

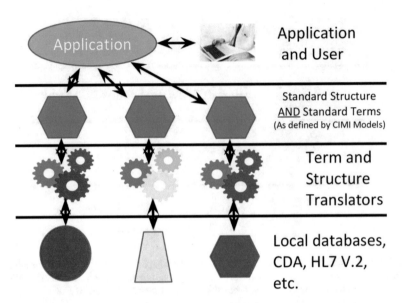

Fig. 5.4 The shapes are all transformed into the preferred 'blue' representation and the terms are also mapped to a standard defined by a CIMI model (using the data standards we will discuss in a later chapter) for each clinical condition. The three EMRs are now semantically interoperable. (Courtesy Dr. Stanley M Huff)

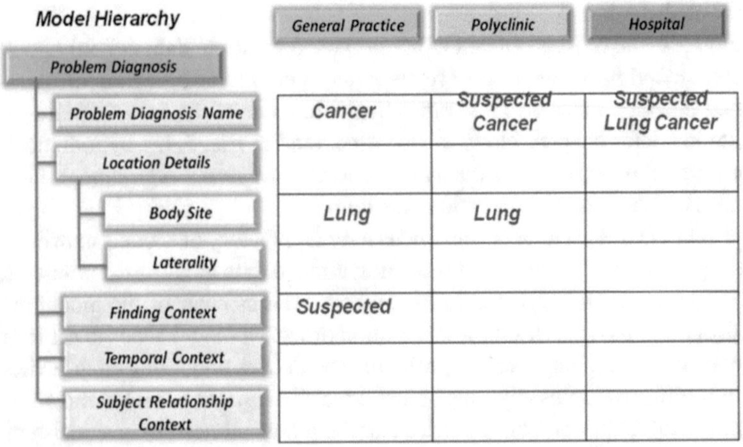

Model Hierarchy	General Practice	Polyclinic	Hospital
Problem Diagnosis			
Problem Diagnosis Name	Cancer	Suspected Cancer	Suspected Lung Cancer
Location Details			
Body Site	Lung	Lung	
Laterality			
Finding Context	Suspected		
Temporal Context			
Subject Relationship Context			

Fig. 5.5 A simple illustration of a CIMI model for a problem/diagnosis begins to show how complex a clinical model becomes if semantic interoperability is the goal. (Courtesy Dr. Linda Bird)

5.4   Applicadia Video

Earlier, we discussed some innovative approaches to physician charting. Now that we have discussed semantic interoperability, you can better appreciate the approach to EHR documentation that Applicadia has developed. Richard Esmond, the company's Chief Technical Officer and co-chair of the CIMI workgroup, has posted a brief video that you should watch on YouTube.[2] In it at around the 50 second point, you will see what he calls a 'graph' that is always focused around a clinical 'concept' node and presents the relationships it has to any other concept nodes known to the platform. We will return to that graph later on.

The node at the focus could be from SNOMED-CT, ICD, LOINC or RxNorm, some of the health data standards we will discuss in a later chapter. The node also might be a synthetic knowledge element that the company has parsed or analytically computed and then attached to that node through a relationship.

Finally, a small, but growing number of nodes are derived from clinical-knowledge artifacts in CIMI Models.

The video, though short, is very rich conceptually and illustrates a number of interesting possibilities:

- The use of voice recognition for medical documentation
- The use of Natural Language Processing (NLP) to properly place voice narration into the traditional sections of a medical note and even the proper body system area of the Review of Systems, a key part of a medical note
- Automatic encoding of free text notes into complex data standards such as SNOMED-CT and ICD-10
- The further use of NLP to properly map voice narrative to the clinical concept graphs thus creating a basis for semantic interoperability
- The use of a concept graph to guide the physician to properly code into the very clinically detailed ICD-10 system. The discussion of the History of Present Illness part of the note concerning the patient's knee problem illustrates this. Similarly, this use of a concept graph could help the physician properly report quality metrics or it could remind them to provide important, but missing, clinical details for a concept they have identified.

You may wish to use this list of capabilities as a guide as you view the video.

5.5   Semantic Interoperability Through Machine Learning

Given its complexity and the long road that achieving it has proven to be, it should not be surprising that there are many points of view on semantic interoperability. Lloyd McKenzie is an internationally recognized expert in HL7 data modeling and

---

[2] https://www.youtube.com/watch?v=INWxXJThIZE

design and an advisor to Canada Health Infoway, an organization with roughly the same role that the ONC plays here in the US. In March 2013, he posted this on Grahame Grieve's blog: "Really robust semantic interoperability is going to be dependent on machine learning and other technologies that will allow the computer to be at least as good at reasoning and inferring context as humans are now – at which point we won't need discrete data at all. Computers will be able to take a big blob of text/dictation/images/etc. and extract all the relevant information needed for their functions without it being neatly divided up for them."[3]

This is an important point of view. Later, we will discuss the potential for machine learning in healthcare. For now, it is worth noting that the explosion in the use of this technology was in large part due to the availability of huge digital datasets on the Internet. After all, the 'learning' part of 'machine learning' comes about when computers process a huge number of examples of something. This might be pictures of cats. It might be examples of some text translated into another language. It also might be EHR notes about rheumatoid arthritis. Those first two datasets have been widely available for quite some time because of the Internet. To be fair, they are also often quite a bit more consistent than EHR notes. Finally, of course, there are no privacy laws covering cat photos or openly available translations of text. Nevertheless, with EHR adoption now widespread, and assuming we can solve the easier levels of interoperability and have broader access to clinical data, might semantic interoperability be best left to machines to figure out?

An early indication that this might be the case is Google's Healthcare API announced by the company at HIMSS 2018. In his opening keynote address at this huge meeting, Eric Schmidt, Technical Advisor and former Chairman of Google's parent company, Alphabet, described a hypothetical future product he called Dr. Liz – named in honor of the first woman to earn a medical degree, Elizabeth Blackwell. This voice assistant in patient rooms would interact with consumers, makes evidence-based recommendations to doctors and largely free them from the burden of working in an EHR.

He said that to achieve this hospitals in the future would need "a clinical data warehouse packed with diverse data sets that are curated and normalized such that sophisticated analytics can be run against the data and accessed with a rich API. Hospitals then need a second tier of data to supplement EHRs."

He went on to say that "EHRs are an incredibly important breakthrough in getting data in place, revenue opportunities, and they manage the workflow of the organization." Apparently referring to the quite intentional omission of semantic interoperability in the FHIR standard, he further said that the "sum of what they do is crucial. But FHIR will not fully get the information out and that's why you'll need the mid-tier data store."[4]

---

[3] http://www.healthintersections.com.au/?p=1407

[4] http://www.healthcareitnews.com/news/eric-schmidt-lays-out-formula-healthcare-innovation

5.6 Interoperability and Meaningful Use

The Meaningful Use criteria are intended to promote robust interoperability between the systems used by patients, providers, and healthcare institutions. The Meaningful Use stages progressively demand compliance with the Patient Engagement Framework (PEF) shown in Fig. 5.6. It was developed by the HIMSS Foundation and the National eHealth Collaborative that are now merged. They focus on better health and greater value through information technology. You can see across the bottom how the stages of MU align with the PEF.

Stage 1 of Meaningful Use, the "Engage Me" phase of the PEF, specifies that patients can view and download their electronic health record. Stage 2, the "Empower Me" phase of the PEF, is far more ambitious and specifies integration with a health information exchange (HIE), the topic of the next chapter, as well as e-referral coordination among providers; ambulatory and hospital records integration; the addition of images and video to EHRs; and the inclusion of data from commercial labs, radiology, and pharmacies. We are currently quite far away from having all of this everywhere but each of these capabilities is available in at least some places.

Presumably, because of Meaningful Use, a New York study showed the reported rate of PHR use increased from 11% in 2012 to 17% in 2013. More directly to the point, the proportion of the PHRs provided by doctors or healthcare organizations

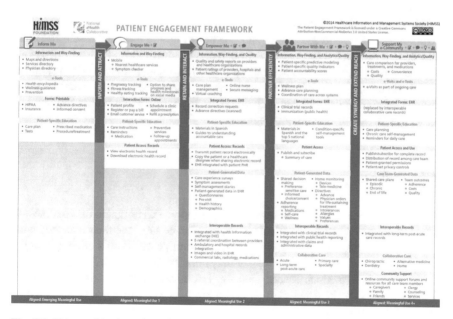

Fig. 5.6 This graphic shows how the Patient Engagement Framework developed by the HIMSS Foundation and the National eHealth Collaborative aligns with the three stages of Meaningful Use.

increased sharply from 50% in 2012 to 73% in 2013. This research also showed that the mean age of PHR users was 47.2 years, 51% were female, and 80% had a physician who used EHRs.[5]

5.7   HIPAA

Later we will discuss health information exchange technologies for sharing health data. First, any transport mechanism for HIE must assure privacy, privacy and trust. *Privacy* means sharing data only with patient permission. *Security* means protecting that data from access by unauthorized entities. To create *Trust,* an entity must assure the identity of others entities with which they share data.

In the US, the Health Insurance Portability and Accountability Act of 1996 (HIPAA) created a Privacy Rule. The government says it "establishes national standards to protect individuals' medical records and other personal health information and applies to health plans, health care clearinghouses, and those health care providers that conduct certain health care transactions electronically."[6] The inclusion of health insurance plans and claims clearinghouses and the focus on electronic transactions such as claims for payment may seem surprising to those focused on clinical use cases but the law also mandates uniform standards for electronic transmission of administrative and financial data relating to patient health information. The intent was to move these transactions from paper to electronic form and that has been quite successful.

The Rule requires appropriate safeguards to protect the privacy of personal health information and sets limits and conditions on the uses and disclosures that may be made of such information without patient authorization. The Rule also gives patients certain rights over their health information, including the rights to examine and obtain a copy of their health records, and to request corrections. HIPAA penalties can be substantial and can include prison time, so staying within the law is a priority at all health care organizations.

5.8   Privacy

We touched on Privacy in our discussion of the Indivo PCHR and in the HealthVault exercise when you saw how patients could specify the subset of their data they wish to share. In practice, patients still often sign a form provided on a clipboard in the waiting room and, although they may or may not read what it says, they are agreeing to share all their medical record data for Treatment, Payment, and Health Care Operations (TPO). In simple terms, this allows sharing of their data with others involved in their treatment such as when their doctor refers them to another

[5] https://www.ncbi.nlm.nih.gov/pmc/articles/PMC4026516/

[6] https://www.hhs.gov/hipaa/for-professionals/privacy/index.html

physician. It also allows use of their data in claims for payment for services rendered and for operational purposes such as quality reporting.

Beyond the routine use encompassed by TPO, there are many 'secondary purposes' for using health data. These include comparative studies, policy assessments, and life science research. These uses of *identifiable* or Protected Health Information (PHI)   require specific permission of the patient. *De-identification* commonly removes data from privacy rule protections by eliminating 18 CMS specified (the Safe Harbor rule fields that could identify the individual patient. *Limited* datasets still contain dates for admission, discharge, and service provision, date of birth, date of death and city, state and zip code. The use of limited datasets for research does *not* require specific patient permission, but it does require that the data be treated the same as PHI since re-identification is possible, given the inclusion of this temporal and demographic information.[7]

To avoid this, in some circumstances, researchers randomly alter dates and other data where this will not compromise its research value. Genomic data presents special challenges because each patient's genome is unique so altering the genomic data can severely compromise its value.

There is a second CMS approved, but less often used, Expert Determination Method The expert may have a statistical, mathematical, or scientific background and certifies that the risk of re-identification is 'very small'.[8]

However, no matter how it is done, de-identification of health data presents challenges as described by the health informatics group within the NORC at the University of Chicago:

Lack of utility: Commonly used de-identification methods such as HIPAA's Safe Harbor can eliminate too much valuable data that have analytic utility

Time Consuming and Error Prone: Manual processes involved in disclosure review of output data are time consuming, labor intensive, and typically lack rigorous and comprehensive quality control

Abundance of Individual Data: Exponential growth of available data on individuals makes frequently employed de-identification methods subject to re-identification disclosure risks

Shortage of Disclosure Experts: There are not enough experienced statistical disclosure limitation experts to meet demand

Disclosure Risk: Can lead to an inability to fully leverage the value of the data (monetary, research value), brand damage, and financial penalties[9]

---

[7] https://www.hhs.gov/hipaa/for-professionals/privacy/special-topics/de-identification/index.html#standard

[8] https://www.hhs.gov/hipaa/for-professionals/privacy/special-topics/de-identification/index.html#guidancedetermination

[9] http://www.norc.org/PDFs/BD-Brochures/PHR%20NORC%20X-ID%20Brochure%20JH%20Version.pdf

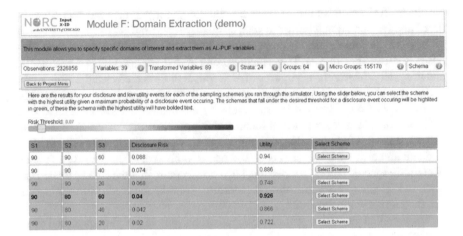

Fig. 5.7 A NORC developed X-ID tool assists an expert in balancing the utility of a health dataset for a specific use case against the risk of re-identification of the data. (Courtesy NORC)

The expert method might be preferable where the Safe Harbor rule makes data unattractive for specific purposes but, as NORC discusses, there is a shortage of qualified experts and it can also involve some manipulation of the data so the NORC group developed X-ID, a tool shown in Fig. 5.7, that uses proprietary statistical methods to give data owners the ability to experiment digitally to define their desired risk and analytic utility thresholds.[10]

5.9 Security

By far the most commonly used technology to secure data is Public Key Encryption (PKI). You use it whenever you visit web sites whose address begins with HTTPS. PKI uses a pair of public and private keys (X.509 certificates) to secure data. In simple terms, a Certificate Authority issues the keys and binds them to the identity of an entity. Before that, a Registration Authority, which could also be the Certificate Authority, verifies that the entity is who they claim to be.

The keys are large numbers with a mathematical relationship.[11] As a result, data encrypted using *either* key can only be unencrypted using the other one. Importantly, the public key is freely shareable but the private key must remain secret. However, as we will now see, there are use cases for using *either* key to encrypt something and the opposite key to decrypt it.

---

[10] http://www.norc.org/Research/Capabilities/Pages/x-id.aspx

[11] The security is provided because the relationship is prohibitively time consuming and expensive to discover using *currently available* computing technologies You may have read that Quantum Computing could be problematic should it prove practical. This could be one of the potential problems if hypothetical quantum computers could overcome the difficulty of discovering the relationship.

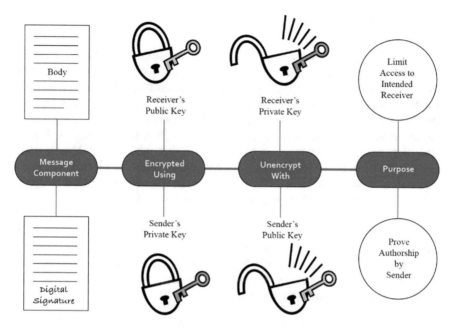

Fig. 5.8 Both the public and private keys are useful in PKI. As shown along the top of this figure the receiver's public key is used to encrypt a document which only the receiver can then unencrypt because their matching private key is securely available only to them. Securing private keys is critical to the protection of the data and must therefore be managed carefully. This is done automatically for users by modern browsers. Along the bottom the sender's private key is used to create a 'digital signature'. Anyone can unencrypt it using that sender's public key but that is fine because the purpose here isn't securing data it is establishing trust in the identity of the sender and assurance that the data wasn't tampered with during transmission. (Author)

In the most common use case, shown along the top of Fig. 5.8, the sender encrypts a document they wish to share using the intended *receiver's public key*. Since only the recipient has the matching private key, only they can decrypt and read the document.

As shown along the bottom of Fig. 5.8, PKI also involves a digital signature. Its purpose is to prevent tampering and impersonation in digital communications. Creating it involves hashing which is the transformation of a string of characters into a usually shorter fixed-length value or key that represents the original string. The hash is then encrypted using the *sender's private* key. The encrypted hash -- along with other information, such as the hashing algorithm -- is the digital signature. The reason for encrypting the hash instead of the entire message or document is that a hash function can convert an arbitrary input into a fixed length value, which is usually much shorter. This saves time since hashing is much faster than signing. Also the fixed length can be taken into account when processing or storing the message digest. A recipient can confirm the source of the document by decrypting the digital signature using the sender's *public* key. Please note that this in no way compromises the sender's private key, it just confirms that the digital signature was encrypted using it. Also, since the document wasn't hashed, this does not compromise confidential patient information.

5.10   Trust

The use of digital signatures is an example of technology to establish trust in the origin of a document. However, trust has no purely technologic solution. Some trusted organization ultimately must verify that entities involved in the exchange of health information are who they claim to be. Hospitals already do this when they credential providers to care for patients in their institution. Establishing trust is harder when patients become involved in adding information to their EHR or in retrieving information from it. For an example of this, think back to HealthVault's support of Direct email for data sharing. As we will discuss next, the DirectTrust framework assures that Direct email addresses belong to the intended recipient. Special Direct servers (HISPs) only accept Direct email addresses creating a closed system.

How might a provider know for sure that such an email they receive actually comes from their patient? This problem is harder when patients receive care from multiple organizations, each of which must trust that the other has properly validated their identity. Because we do not have a universal patient ID here in the U.S., even being sure of the identity of patients can be challenging.

5.11   Blockchain in Healthcare

At present, few areas of technology are creating more discussion than blockchain (also known as Distributed Ledger Technology or DLT) and this definitely extends to healthcare and electronic medical records in particular. An interesting 2016 paper from MIT's MediaLab and Beth Israel Deaconess Medical Center concludes that "principles of decentralization and blockchain architectures could contribute to secure, interoperable EHR systems".[12] One of the authors of this paper is John D Halamka, MD who is CIO of Beth Israel Deaconess Medical Center, a practicing emergency physician, Healthcare Innovation Professor at Harvard Medical School and Chairman of the New England Healthcare Exchange Network. Previously he convened The Argonaut Project that created tools for the practical application of FHIR to Meaningful use. He has subsequently become Editor in Chief of the online journal *Blockchain in Healthcare Today*.[13] Finally, in a March 2018 post on his blog, Grahame Grieve discusses some possible use cases for blockchain in healthcare and how HL7's standards efforts might need to consider it in the future.[14]

For our purposes, we will think of a blockchain as a decentralized digital shared ledger or record book with each line (actually each block) containing groups, or "blocks," of transactions representing all new additions to the blockchain crypto-

---

[12] http://dci.mit.edu/assets/papers/eckblaw.pdf

[13] https://blockchainhealthcaretoday.com/index.php/journal

[14] http://www.healthintersections.com.au/?m=201803

graphically "chained" together. However, unlike a physical ledger, there are multiple copies stored on computers all around the world in multiple locations and each of them is able to validate and verify the entire blockchain. (i.e., to verify that the copy on any of them is the same as on as all the others making it virtually impossible to make illicit changes without being detected). Distributed Ledger Technology (DLT) can record many kinds of things (e.g., cryptocurrencies like Bitcoin and Etherium).

The issue of interest here is how to use DLT technology in healthcare. People often assume that actual patient records will be stored in the blockchain, but that may not be practical or scalable given the sensitivity and the shear amount of data involved. Some health insurance companies are piloting it as a way of having up to date information on providers.[15] Some new efforts propose to use it to essentially create a uniform patient ID. Finally, as the FHIR standard gains traction, it might be possible to take advantage of its granularity to store patients' specific permissions for data sharing in a blockchain.

To summarize, at present there are many types of blockchains and many use cases for them in healthcare at various stages of development. To consider how blockchain might be in healthcare we will discuss its two main categories. **Permissionless**, or trustless, blockchains support anonymous transactions of cryptocurrencies and/or other digital assets where the parties are anonymous but there is complete public transparency as to the transaction itself. In other words, in a Bitcoin transaction the funds being moved are clear but the identity of the parties to that transaction is not available. The other **permissioned** type is, in some ways, the polar opposite for use in applications where the identity of the parties must be unequivocally known to one another but the actual information must be concealed. Clearly, healthcare applications fall more naturally into the later permissioned category where the various "validating" nodes are known and trusted by one another.

A newer category involves the concepts of **self-sovereign identity** and smart contracts they could be used to extend a health system's pre-existing trust in providers and patients beyond its traditional boundaries. To give two examples of this, patients using FHIR apps could attest to their own identity to multiple EHRs in order to gain access to their records or they could self-manage consent to the sharing of their clinical data.

5.12   Health Information Exchange: Direct

A health information exchange (HIE) is a specialized network for the secure sharing of patient data. There are various HIE architectures that correspond functionally to the three forms of interoperability we discussed earlier.

---

[15] https://www.businesswire.com/news/home/20180402005181/en/Humana-MultiPlan-Optum-Quest-Diagnostics-UnitedHealthcare-Launch?elq_mid=10941&elq_cid=558581

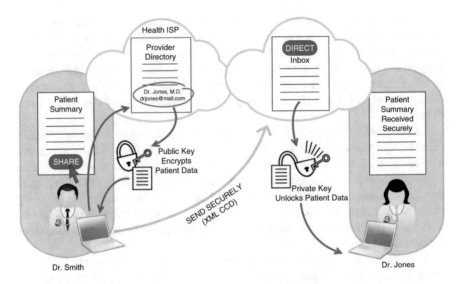

Fig. 5.9 A simplified illustration of how Direct uses email and PKI to securely send a patient summary (in the form of an XML formatted CCD) as an encrypted email attachment from Dr. Smith, the patient's PCP, to Dr. Jones, a specialist to whom he wishes to refer that patient. (Author)

You should recall that the first layer of interoperability is simple transport of the data from one provider to another. The transport mechanism may have no knowledge of or understanding of the data. ONC created the Direct Project in 2010 to specify a simple, secure, scalable, standards-based way to send authenticated, encrypted health information directly to known, trusted recipients using secure email. Figure 5.9 is a graphic illustrating Direct. For simplification it assumes, among other things, that both providers are using the same Direct server, called a Health ISP[16] or HISP. Direct builds on existing technologies for email that we will discuss next and for securing data (such as PKI). Most EHR, HIE and PHR vendors, including HealthVault, support Direct. When you did the HealthVault exercise you should have seen that your account automatically provides a special Direct email address that can be used to securely send and receive health data but only to or from other Direct email addresses. In some EHRs, providers can initiate Direct messages from within a charting session, such as when they are referring a patient to another provider.

We will now briefly discuss some of the Internet and email standards and processes used by Direct HISPs. Non-technical readers may wish to skip ahead.

- To assure secure transport Direct combines the Simple Mail Transfer Protocol (SMTP) with the special Direct e-mail addresses we just discussed and associated with X.509 certificates.

---

[16] The acronym ISP stands for Internet Service Provider and it is 'borrowed' by Direct for what is a specialized email service and not the full Internet service we are all familiar with.

- In most cases, an entity wishing to have an email address and credentials issued goes through a process documented in a "trust framework" established by DirectTrust, a collaborative non-profit association of health IT and health care provider organizations. This trust framework supports both provider-to-provider and bi-directional exchange between consumers/patients and their providers.
- Direct uses Secure/Multipurpose Internet Mail Extensions (S/MIME), a public key standard for encrypting email attachments that can be in virtually any format from a fax image to a PDF to a structured, XML-formatted HL7 C-CDA document.
- Direct uses encrypted and signed Message Disposition Notifications (MDN) to confirm delivery. This can be particularly important in situations such as when a lab sends a critical result back to the ordering provider using Direct.
- Domain Name Servers (DNS), the Internet's equivalent of a phone book, maintain a directory of domain names and translate them to Internet Protocol (IP) addresses. Lightweight Directory Access Protocol (LDAP) is an Internet protocol that email and other programs use to look up information from a server. Direct uses both DNS and LDAP to discover the certificates of message recipients prior to sending a message, in order to fulfill the encryption functions of S/MIME.

5.13   Health Information Exchange: HL7 Messaging

The second layer of interoperability we identified earlier provides a structure to contain data elements so that the receiving system knows what each element represents, even though it may have no semantic understanding of the contents. A common example is electronic lab reporting of results back to the provider who ordered a test. According to the CDC in 2013 around 90% of all laboratory test results reports from the over 200,000 certified laboratories in the US used HL7 messaging.[17] There were still issues with this because only 29% used the newer V2.5.1 standard required by Meaningful Use. However, that notwithstanding, this is clearly one of the most electronic areas of healthcare. As of February 2014 patients have the right to direct access to these results.[18] At least one study showed increased patient engagement as a result of direct access to laboratory results but also increased anxiety as a result of not understanding the results and their ramifications possibly leading to more visits.[19]

Earlier, we mentioned that HL7's messaging standard dates from the late 1980's. The Version 2 standard, shown in Fig. 5.10, is based on EDI/X12 technology from that era. EDI was designed to be read only by computers and is both dense and cryptic to save space in the limited storage of the computers of that era.

---

[17] https://www.cdc.gov/elr/elr-meaningful-use.html

[18] https://www.federalregister.gov/documents/2014/02/06/2014-02280/clia-program-and-hipaa-privacy-rule-patients-access-to-test-reports

[19] https://www.ncbi.nlm.nih.gov/pmc/articles/PMC4919031/

```
MSH|^~\&|LABGL1||DMCRES||199812300100||ORU^R01|LABGL1199510221838581|P|2.3
    |||NE|NE
PID|||6910828^Y^C8||Newman^Alfred^E||19720812|M||W|25 Centscheap Ave^^
    Whatmeworry^UT^85201^^P||(555)777-6666|(444)677-7777||M||773789090
OBR||110801^LABGL|387209373^DMCRES|18768-2^CELL COUNTS+DIFFERENTIAL TESTS
    (COMPOSITE)^LN|||199812292128||35^ML|||||||
    IN2973^Schadow^Gunther^^^^MD^UPIN
    |||||||||^Once||||||CA20837^Spinosa^John^^^^MD^UPIN

OBX||NM|4544-3^HEMATOCRIT (AUTOMATED)^LN||45||39-49
    ||||F|||199812292128||CA20837
OBX||NM|789-8^ERYTHROCYTES COUNT (AUTOMATED)^LN||4.94|10*12/mm3
    |4.30-5.90|||F|||199812292128||CA20837
```

Fig. 5.10 An electronic laboratory test result report in the HL7 V2 messaging format. Each line begins with a three letter 'Segment ID'. In this example the OBR segment is the laboratory test order and, among other things, it contains the name of the test in text, the name of the physician who ordered it and the LOINC code for the test (18768-2). The OBX segments are the results and the normal ranges that you should be able to use to verify that both results are normal. (Courtesy HL7)

We also said earlier that the OBX lines (segments) are test results – the patient's hematocrit and erythrocyte count – both indicators of anemia, if reduced. Study of the two lines should show you that both are normal in this patient but this older EDI format is definitely not intended for human consumption!

The OBR segment is the test order. If you search for the number 18768-2 from that line you should easily confirm that it's the LOINC code for this test. This is an example of the use of data standards within a messaging standard. LOINC, if it were used consistently, could create semantic interoperability for that one data item but unfortunately it isn't used consistently and adoption of LOINC codes by both laboratories and providers remains a challenge in part because of the complexity of the coding system that we will discuss later.[20] The newer V3 HL7 messaging standard uses XML which is more human readable. However, as illustrated by the CDC statistics for electronic lab reporting, V2 is so deeply embedded into many existing systems that V3 is not widely used.

5.14   Health Information Exchange: Semantic Interoperability

We discussed semantic interoperability earlier. If the systems sharing data don't support it, a sophisticated form of HIE could create semantic interoperability thereby bridging the many ways the same concepts can be expressed by providers and represented in EHRs and other clinical systems. This involves a complex process that maps external representations of clinical concepts to some standard developed or adopted by the HIE.

---

[20] https://www.fda.gov/downloads/MedicalDevices/NewsEvents/WorkshopsConferences/UCM530349.pdf

Fig. 5.11   Recent IHIE participation statistics that apparently update in real time on the HIE's web home page. At this time 117 hospitals from 38 health systems; over 42,000 providers in nearly 15,000 practices; and nearly 13.5 million patients had access to over 11 *billion* clinical data elements (with access permission of course). (Courtesy IHIE)

The premier example of this here in the US is the Indiana Health Information Exchange (IHIE [pronounced EYE-HIGH]) that says it is the largest such organization in the country. This claim is easy to believe based on the participation statistics (Fig. 5.11) that proudly greet visitors to the IHIE site.[21]

Figure 5.12 presents a high-level diagram of the IHIE technical architecture. The data governance box is central, as it should be, since it is where the curation of data from the many disparate sources shown on the left occurs to establish semantic interoperability. On the right, you can see how this governance enables a number of value-added services. These include the reporting of laboratory test results and of the quality metrics that are increasingly required for providers under the new payment models we also discussed earlier.

The Regenstrief Institute in Indiana created the expensive and sophisticated technology used by IHIE. Support for it came from The Regenstrief Foundation, a philanthropic organization that describes its mission as "to bring to the practice of medicine the most modern scientific advances from engineering, business, and the social sciences, and to foster the rapid dissemination into medical practice of the new knowledge created by research." Absent such an unfortunately rare funding source, this type of HIE is usually economically impossible to create.

In the case of IHIE as you see in Fig. 5.12, all the curated data is aggregated and stored centrally. This architecture is convenient for data governance, analysis and reporting. However, it makes buy-in difficult for organizations that might be leery of turning over their data and concerned about how it may be used and how their performance might look. Again, Indiana is a special case, where, over many years, IHIE has gained the trust of the many care organizations involved in the exchange.

---

[21] http://www.ihie.org/

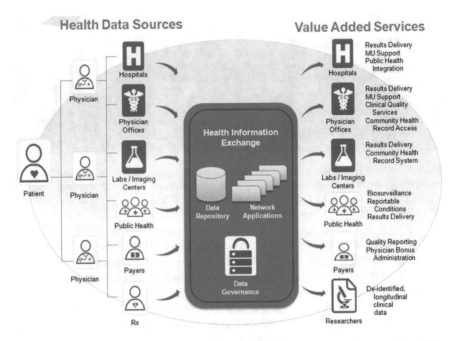

Fig. 5.12 This high-level diagram of the IHIE architecture centers as it should around the HIE's sophisticated Data Governance engine that essentially creates semantic interoperability among and between the many data sources (on the left) in order to enable not just data sharing but the value added services shown to the right. (Courtesy IHIE)

5.15   The Federated Model

This level of trust in a shared, centralized architecture is rare. The concept of a 'federated' HIE goes a long way toward resolving these concerns. In this model, the data stays at the source. There is a standard for queries and responses and, based on the query, participating organizations usually have the option of responding or not. In a federated model, either the nodes must translate their data into some standard response format or they must map their information into a standard data model usually stored on a separate server that then serves to handle queries and responses.

By far the most successful example of this at present is the Observational Health Data Sciences and Informatics program (or OHDSI, pronounced "Odyssey") and its OMOP data model.[22] OHDSI was created to enable active drug safety surveillance with government and pharmaceutical industry support using observational data collected during patient care. There is a brief entertaining video explaining this on the OHDSI site.[23] On that site there is also a map illustrating OHDSI's global reach and listing its collaborators.[24]

---

[22] https://www.ncbi.nlm.nih.gov/pmc/articles/PMC4815923/

[23] https://ohdsi.org/who-we-are

[24] https://ohdsi.org/who-we-are/collaborators/

The OMOP data model is the engine that drives OHDSI. All collaborators map their clinical data to OMOP establishing a form of semantic interoperability. The current version of OMOP is 5.2 and it is publicly documented on a GITHUB Wiki.[25] The core components of the specification are 12 Standardized Vocabularies and 14 Clinical Data Tables. The tables provide the structure and the vocabularies specify its contents – the two requirements for semantic interoperability. Interestingly OMOP concepts can be either the public SNOMED-CT or RxNorm standards we will discuss later or are "custom generated to standardize aspects of observational data". Similarly, the vocabularies are "collected from various sources or created de novo by the OMOP community".

Each of the OHDSI collaborators is available to respond to queries from any other collaborator. Often the use case is seeking specific patient cohorts for clinical drug trials. To facilitate this OHDSI provides three analytic and exploratory tools for data in the OMOP model[26]:

ATLAS – a web-based integrated platform for database exploration, standardized vocabulary browsing, cohort definition, and population-level analysis
ACHILLES – a standardized database profiling tool for database characterization and data quality assessment
CALYPSO – an analytical component for clinical study feasibility assessment:

It also provides two tools for exploring other data sets of potential interest to OHDSI collaborators:

KNOWLEDGEBASEWEB – an experimental user interface for exploration of data present in the Largescale Adverse Effects Related to Treatment Evidence Standardization (LAERTES) evidence base that provides data from a wide variety of sources with information relevant for assessing associations between drugs and health outcomes.
Drug Exposure Explorer – visualize drug exposures (at the time of this writing it was an experimental deployment using the huge CMS SynPUF simulated patient dataset.)[27]

The history of this effort is interesting. Earlier we talked about the Indiana Health Information Exchange (IHIE) and the Regenstrief Institute. Regenstrief made two core contributions to OMOP and OHDSI. The first was that Marc Overhage, MD, PhD was a principal investigator on the original grant that built OMOP. He was also a key contributor to the community wide electronic medical record (the Indiana Network for Patient Care) containing data from many sources including laboratories, pharmacies and hospitals throughout Indiana. The system connects nearly all acute care hospitals in the state and includes inpatient and outpatient encounter data, laboratory results, immunization data and other selected data. He also helped create IHIE

---

[25] https://github.com/OHDSI/CommonDataModel/wiki

[26] https://www.ohdsi.org/analytic-tools/

[27] https://www.cms.gov/Research-Statistics-Data-and-Systems/Downloadable-Public-Use-Files/SynPUFs/DE_Syn_PUF.html

Jon Duke, MD is now Director of the Center for Health Analytics and Informatics at Georgia Tech Research Institute and Principal Research Scientist at Georgia Tech College. He was a lead developer of the OHDSI software and was involved in some of the studies done using it. The diversity of the studies is interesting. They range from elucidating treatment pathways for millions of patients with type 2 diabetes mellitus, hypertension, and depression[28] to the impact of birth month on future disease risks[29] to a more traditional study done using OHDSI of the increased risks associated with taking certain medications.[30] We will look in detail at two research projects done using OHDSI in the final chapter of the book.

Finally, the IHIE was one of the first data sites to connect to OHDSI and it remains IHIE's only HIE data site.

5.16   The Utah Health Information Network

The Utah Health Information Network (UHIN) was founded in 1993 by area hospitals, healthcare providers, health plans and governmental stakeholders to offer a non-profit, secure, standards-based approach to exchange health insurance claims information. Divergent standards and procedures made electronic claim submissions inefficient and costly. The creation of UHIN's clearinghouse goal was to overcome these obstacles. It was apparently successful because today UHIN's clearinghouse offers service to 90% of Utah's providers and other customers in 42 states. Annually UHIN processes in excess of 36 million claims/payment remittances, 24 million remittance advices, 36 million eligibility requests, and over a billion HIPAA transactions.

In 2007 UHIN's Board of Directors charged the organization with creating a clinical health information exchange (CHIE) comprised of a master patient index (MPI) and record locator services (RLS) through which it could provide a longitudinal patient medical record view and a central (in hospitals the term clinical is often used) data repository (CDR).

Today the CHIE houses over 53 million records on five million unique patients from Utah and around the world. The CHIE is now connected to over 90% of the hospitals in Utah, with a goal of achieving universal connection by the end of 2018. Labs, clinics and long-term care facilities across the state are also connected. Additionally, as part of the western implementation of the Patient Centered Data Home, spearheaded by the Strategic Health Information Exchange Collaborative (SHIEC), UHIN is fast becoming a western hub for data exchange between states. UHIN is connected to HIEs in six additional western states. Moreover, UHIN is also connected with hospital systems in two neighboring states that lack an HIE.

---

[28] http://www.pnas.org/content/113/27/7329.short

[29] https://academic.oup.com/jamia/article/22/5/1042/930268

[30] http://onlinelibrary.wiley.com/doi/10.1111/epi.13828/full

Data sources such as hospitals, clinics and labs send patient data to UHIN's interface engine according to the type of connection the source has with UHIN. Once received, the data flows to the MPI engine, where it is attributed to the correct patient. Full names, birthdates, "nicknames," addresses and other variables are cross checked to ensure correct attribution. The data then flows to a data curation layer, where specific data elements are mapped to standard formats and terms, and afterwards sent to the central data repository.

The data found in the CDR then undergoes a process where standardized terminologies (e.g. LOINC, ICD-10 and SNOMED-CT) are added to the data and unstructured data are passed through a Natural Language Processing (NLP) engine before being sent to UHIN's Data Warehouse.

UHIN members access data on the specific patients with whom they have a HIPAA-approved treatment relationship through a web-based portal. As shown in Fig. 5.13, providers can obtain analytical information, dashboards and other reports. Physicians using this interactive visual analytics report can see their patients' information across various hospitals. On the right are filters and clicking on a data set will reset the graph to focus on that selected data set.

Figure 5.14 presents the same data geographically by meshing it with geographic information obtained by UHIN from other sources. Patient location can be a key determinant of health and health outcomes. Factors with potential impact on disease development and management include access to quality food or exposure to environmental risks.

The hospital's clinical data repository can also export submitted data related to patients' admissions, discharges or transfers (ADT) to UHIN's "Alerts Engine." Subscribing members receive pushed ADT alerts on their patients. The alerts notify

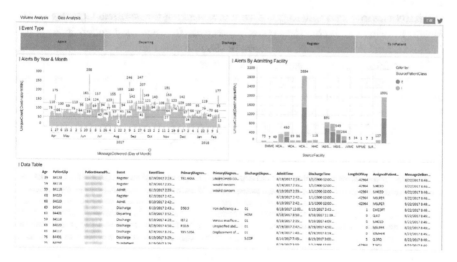

Fig. 5.13 An example of using visual analytics to help providers understand their patients across care settings. A physician can see their patients' information across various hospitals. On the right are filters and clicking on a data set will reset the graph to focus on that selected data set. (Courtesy UHIN)

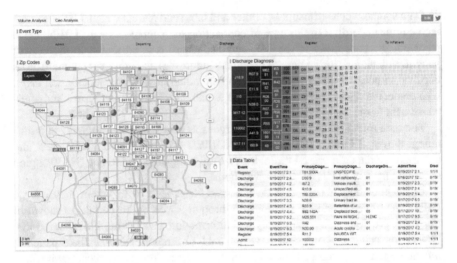

Fig. 5.14 An example of using external geocoded data to put clinical data into context. Patient location can be a determinant of disease risk and outcomes. A common example is asthma and its relationship to air quality. (Courtesy UHIN)

primary care physicians, case managers and others of their patient's status, allowing them to better coordinate care, provide needed follow up care, and proactively reach out to vulnerable patients.

HIEs can provide added value to a variety of stakeholders. An interesting UHIN example is that public health departments can use data from the HIE to monitor chronic disease, to receive physician-provided data for various registries and monitor surveillance for outbreak of disease.

UHIN was awarded a 2015 grant from the Office of the National Coordinator for Health Information Technology (ONC) to expand to previously underserved populations in rural communities, long-term post-acute care facilities, behavioral health, poison control, emergency services, and patients.

To provide patients with access to their own records, UHIN created a portal that allows patients to view all their records from their various providers in one location. Patients can also communicate with their providers through secure email, and they can attach their continuity of care document (CCD). Future enhancements of the patient portal may include access for family members, and the ability to upload information from personal health devices.

UHIN is a pioneer among HIEs in offering support for FHIR that is now one of several ways to send claim attachments, which UHIN says results in a more automated process thus reducing costs. Payers can use the FHIR query to request and receive needed clinical information to process claims or to provide prior authorization for expensive or unusual services. UHIN can respond from the clinical data contained in the HIE and if necessary transmit the request directly to the provider. The provider can respond with their preferred technology, such as FHIR, sFTP, or SOA (refer to the glossary for definitions of these terms) and UHIN will return the

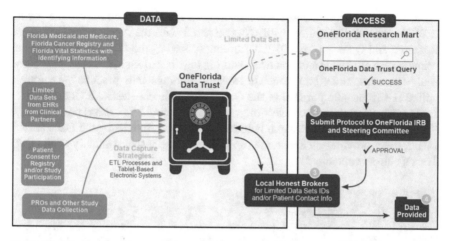

Fig. 5.15 OneFlorida Clinical Research Consortium's architecture is similar to IHIE in that it aggregates and curates data from diverse sources across the state. Unlike IHIE, its sole purpose is community based clinical research. (Courtesy OneFlorida)

request to the payer in their preferred technology, which will likely increasingly be FHIR over time. UHIN has also added a FHIR connection to its provider directory. UHIN's provider directory includes demographic information on the provider including their Direct secure messaging address. UHIN's future use of FHIR may include intelligent alerts, patient portal connections and the exchange of care plans.

5.17   The OneFlorida Clinical Research Consortium

The OneFlorida Clinical Research Consortium is an interesting example of a curated centralized health information data collaborative. Its architecture is shown in Fig. 5.15 and it centers on the OneFlorida Data Trust. According to the consortium's web site it "contains claims and encounter data from Florida Medicaid and Capital Health Plan and robust patient-level electronic health record data from public and private health care systems that are consortium partners. The data includes diagnoses, procedures, medications, patient demographics, unique patient codes for re-identification by consortium partners along with other data elements.[31] It is mapped to the PCORnet Common Data Model (CDM)." The PCORnet site provides a good brief video explaining the purpose of the data model.[32] PCORnet CDM is, in turn, based on the data model developed by the FDA for its Sentinel National Medical Product Monitoring System (Sentinel), another good example of the federated approach that supports remote queries but in which all data remains at the source under the control of the owner of the data.[33]

---

[31] http://onefloridaconsortium.org/about/infrastructure/

[32] http://www.pcornet.org/pcornet-common-data-model/

[33] https://www.sentinelinitiative.org/

The Data Trust complements the consortium's Practice-Based Research Network (PBRN), one of 178 national PBRNs registered with the Agency for Healthcare Research and Quality (AHRQ) [34] According to AHRQ the PBRNs are "groups of primary care clinicians and practices working together to answer community-based health care questions and translate research findings into practice." Therefore, unlike IHIE, the sole purpose of the OneFlorida Clinical Research Consortium is facilitating research on the state's diverse resident population. It has posted a page listing a wide variety of research publications derived from using the Data Trust.[35] Like IHIE it provides an apparently up-to-the-minute visual summary of the Data Trust's current contents.[36]

Case Study: Surescripts®

Pharmacy was one of the early implementations of computers in healthcare. Pharmacists used typewriters that were easily replaced by a keyboard. The needed labels and other paper documents and the pharmacy transaction itself was focused and relatively simple as compared to a medical record. As a result, there was a clear business case for pharmacy automation. While most of the early use of computers in healthcare was inside the hospital the use of computers in community pharmacies was virtually universal by the end of the twentieth century. However, these systems largely lacked interoperability between pharmacies (except in the case of chains where Walgreens in particular was early to establish electronic connection among their pharmacies). This introduced the potential for pharmacists to have an incomplete view of patient's medications with a resulting inability to verify that they did not duplicate or interact with other medications. This also reduced the pharmacist's ability to detect abuse of medications such as opioids.

Specialized Pharmacy Benefit Management companies (PBMs) that offered pharmacy benefit plans to employers and others came into existence starting in 1968 when Pharmaceutical Card System Inc. (PCS, later AdvancePCS) was founded. PCS began as a simple plastic benefit card that offered employees discounts at their participating community pharmacy. As medication costs increased, PBMs grew in size and importance and their role expanded to managing pharmaceutical cost through various means including a "formulary" of allowed medications that often gave preference to "generic" medications that came available after the original manufacturer's patents expired. Managing an increasingly complex benefit without a significant electronic connection between the PBM and the pharmacy was problematic.

(continued)

---

[34] https://pbrn.ahrq.gov/

[35] http://onefloridaconsortium.org/research/publications/

[36] http://onefloridaconsortium.org/data/

A 2014 review 0f 47 articles concluded that replacing paper prescriptions with electronic "e-prescribing" "reduces prescribing errors, increases efficiency, and helps to save on healthcare costs".[37] This, of course, requires an electronic means of connecting the physician's EHR with the pharmacy's systems. As a result of its importance, e-prescribing is a goal of the Meaningful Use program.

By 2001 as a result of these problems, the different areas of the industry were recognizing an urgent need for a technologic solution. The major chain and independent pharmacy associations (National Association of Chain Drug Stores and National Community Pharmacists Association) formed SureScript Systems to create a link to physicians and replace paper prescriptions with e-prescribing. The three largest PBMs – Caremark, Express Scripts and Medco Health – formed RxHub to create a link between payers and prescribers.

In 2008 these competing organizations combined their efforts as Surescripts. The goal was to build a national network to connect clinicians, EHRs, hospitals, PBMs, pharmacies and technology vendors in order to maximize the value of having comprehensive patient information at the point of care. Surescripts operates as a nationwide health information network with ownership split between the associations and PBMs. Today Surescripts is connected to virtually all US pharmacies, PBMs, EHRs and clinicians.

Given that it's central business is interoperability it should not be surprising that it was an early adopter of FHIR. While much of FHIR activity is focused on information access from EHRs in the form of FHIR resources there is an alternate paradigm in which a group of resources along with a header creates the FHIR equivalent of an HL7 message.[38] Surescripts uses this approach to enable communications between PBMs and providers as they order new medications.

While Surescripts operates a nationwide health information network and provides many other services we will focus on the particular use case of a provider prescribing a medication in what follows. The goals are to help providers know what medications each patient is eligible to receive under their health insurance plan, help them prescribe more safely, and monitor their patients' adherence to the prescription directions.

Surescripts provides participating EHR vendors with two technology tools for use in communicating with the providers within their workflow. The most commonly used at present is based on an iFrame, a commonly used method of embedding an area within a web page that is typically used by a server other than the one that produced the web page to present information. For example, this could be a video or a web advertisement.

(continued)

---

[37] http://perspectives.ahima.org/electronic-prescribing-improving-the-efficiency-and-accuracy-of-prescribing-in-the-ambulatory-care-setting-2/

[38] We are getting ahead of ourselves here but Surescripts supports the FHIR Observation, Practitioner, Organization, Patient, Contraindications and Medication resources.

Surescripts also supports the SMART on FHIR platform and its CDS Hooks specification that we will discuss in detail later on but this is not yet used by any EHR vendor in large part because Surescripts developed an alternative way of accomplishing a similar degree of EHR integration well before SMART on FHIR and CDS Hooks were developed.

Surescripts feeds this iFrame a patient specific medication "alert." The specific alert selection is configurable based on business rules. Surescripts obtains the information to drive this display from the patient's PBM which has a complete view of their medications and the dispensing histories for each of them. For example, the alert might be the medication this patient is taking with the poorest compliance based on the percentage of the medication that they should have taken over a time period that they actually obtained from any pharmacy. This is calculated as a proportion of days covered (PDC) metric indicating, for example, that the patient obtained only a 45-day supply over a 90-day period. The provider can click on this alert and obtain more information such as is shown in Fig. 5.16.

You will note in this figure that medication adherence provider alerts are conveniently grouped by the disease for which the medications are prescribed. This could be helpful if the provider is asking for more details because the patient's diabetes or hypertension is not well controlled and has no reason to be concerned about compliance for other medications.

Other medication alerts include:

Missing Medication: where a patient's diagnoses suggest that they should be treated with a particular as yet un-prescribed medication.

High Risk Medication: where a patient's clinical data suggest that the use of a prescribed medication could create complication such as interactions with other medications or other clinical problems.

Finally, in order to assure that its messages are responsive the processing is not performed in real time. EHR vendors can send providers' scheduled patients in advance to Surescripts which passing it on to the PBMs to do their analysis (including determining each patient's eligibility). The results are sent back to Surescripts for later posting in the EHR "area" (iFrame) it owns wherever in the careflow that particular EHR vendor has chosen to present it. All of this is done using FHIR messaging as described earlier.

Later we will discuss the importance of medication reconciliation, the process of assuring that each provider has a complete and accurate record of their patients' medications. Surescripts is planning a medication profile service to support reconciliation.

Fig. 5.16 The message on the left alerts the provider to a specific medication adherence problem in which the patient may have ceased to take the medication (in this case a medication which frequently causes side effects patients do not tolerate). Note that the provider can indicate to the system whether they have confirmed the problem to avoid unnecessary future alerts. The area on the right conveniently shows the PDC percentage by medication and groups medications by the disease for which they are prescribed.

5.18   CommonWell Health Alliance®

CommonWell Health Alliance describes itself as "a not-for-profit trade association of health IT companies working together to create universal access to health data nationwide." Its members represent two-thirds of the hospital EHR market and more than one-third of the ambulatory care EHR market, as well as health IT companies in a variety of other sectors including pharmacy, oncology, and population health. The goal is to build a form of health information exchange connecting the commercial products of its members and offer it broadly at a reasonable cost.

CommonWell's extensive service offerings in support of health information exchange include these key core HIE services:

Patient Enrollment: Enable each individual to be registered and uniquely identified across the entire CommonWell network of diverse commercial systems.

Patient Identification and Linking: Link each individual's clinical records across the care continuum (of CommonWell enabled systems)

Record Locator Service: Create an index of the available locations for each patient's information.

Data Query and Retrieval: Enable caregivers to search, potentially select and receive needed data across the CommonWell network

CommonWell was early to recognize the value of FHIR and began using it for patient identity management in 2014. Just as this book was nearing completion

CommonWell announced it has completed building The Argonaut Project's FHIR specifications into each of its core services (often other technologies such as the older Integrating the Healthcare Enterprise (IHE) specification are also supported so read what follows after this list to see more of how FHIR is used).[39]

- FHIR messages can prepopulate CommonWell with patient demographic and encounter data from each connected system.
- Once links have been created, connected systems can request a list of available remote patient content using queries.
- CommonWell uses the flows defined by the XCA specification to distribute or "fan out" ITI-38 queries to all known remote systems where content for the patient might be available. CommonWell Members will be able to use FHIR-based outbound query and retrieve capabilities to access data across the network.
- A combination of Public Key Infrastructure (PKI), JSON Web Tokens (JWT), and SAML 2.0 (a standard for logging onto systems) are used to authenticate and authorize a given edge system to the CommonWell services.

It says that this makes CommonWell "the first national network to enable comprehensive FHIR-based-exchange at scale". Brightree, a provider of cloud-based software to the home and durable medical equipment and home health & hospice companies, will be the first CommonWell member to deploy a FHIR-based capability to their clients.

CommonWell says it believes that offering FHIR "makes it simpler and faster for technology innovators to exchange health data" because of its "modern architecture which makes it easier for developers – both inside and outside the EHR industry – to connect to the information they are trying to access. Additionally, it creates the stepping stones for more widespread sharing of discrete segments of data, as opposed to the comprehensive summary of care documents that are shared today."

A use case that illustrates CommonWell's vision would be a query of a healthcare data network for a patient's allergies that pulls the information from all locations where the patient has received care and correlates the data into a single list of all that patient's active allergies. To achieve objectives like this, as shown in Fig. 5.17, CommonWell's Document Locator Service supports the FHIR DocumentReference resource and corresponding OAuth2 security specifications for Document Responders, as profiled by the Argonaut Project. CommonWell will also utilize and implement Argonaut specifications for provider directories.

MEDITECH is one of the largest suppliers of healthcare enterprise wide information systems and was an early CommonWell member. The company was the first enterprise software CommonWell member to implement the sending and receiving of C-CDA documents using the FHIR DocumentReference resource, as described above. This effectively provides a community (or potentially nation) wide method for sharing of clinical documents with a MEDITECH health system. The company says that it is also supporting the access by patients to their record (containing data as specified by the Common Clinical Data Set (CCDS) we discussed earlier) via any

---

[39] We have not covered the older IHE XCA specifications for cross EHR data sharing but interested readers can refer to the XCA web site (https://wiki.ihe.net/) for more details.

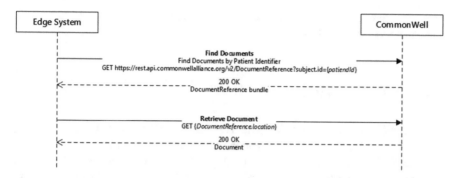

Fig. 5.17 CommonWell's Document Locator Service has implemented a find documents service based on the FHIR DocumentReference resource. The retrieved document could be in the C-CDA format or it could be a PDF or even a FHIR document that consists of a bundle of FHIR resources. (Courtesy CommonWell)

FHIR app of their choosing, a requirement of Meaningful Use. The company further says it is working to provide granular data as FHIR resources and it is also working to implement SMART on FHIR (including support for CDS Hooks, a technique to automatically provide clinical decision support that we will describe later) and will release these capabilities along with a developer sandbox and app gallery to support third party developers.

CommonWell is bridging these specifications to the IHE Cross-Community Access (XCA) specifications that many major EHR vendors already support and are widely used by hospitals and health information exchanges.

5.19 Georgia Tech's Health Data Analytics Platform (HDAP)

Georgia Tech Research Institute (GTRI) constructed the Health Data Analytics Platform (HDAP) by overlaying a FHIR server on the OMOP data model. This provides the advantages of being able to map diverse datasets of interest in the standard data model and having access to that data both via FHIR Resources and using the OHDSI tools. At present FHIR is not well suited to population level analytics (as we will discuss later on, this will likely change in the future) and the OHDSI tools are not designed for granular access to a patient's data so the two technologies are synergistic.[40]

5.20 Data Lockers

While we will not discuss it in any detail here, there is also a model that blends the centralized and federated approaches. In it, all data is stored centrally but in so-called 'data lockers' that remain under the control of the entity that is the source of

---

[40] http://www.hdap.gatech.edu/

the data. That entity typically has the same control of the use of the data they would have in a federated model.

5.21   HIE Challenges

The federal HITECH program supported the development of state HIEs. As a result of that, the growing EHR adoption, and the data sharing requirements of Meaningful Use the 2015 edition of a survey sponsored by the Robert Wood Johnson Foundation reported that, as of 2014, 76% of hospitals were exchanging data with outside health professionals, including ambulatory health professionals and other hospitals. This was up from 62% in 2013 to 41% in 2008. The shared data might include laboratory results, radiology reports, clinical care summaries or medications.[41]

Despite this growth, HIEs reported facing many barriers. Only 46% of operational HIEs and 38% of HIE efforts in the planning stage reported that they were able to cover operating costs with revenue from participants. This has long been an ongoing issue with HIEs so we will now consider how technology might make them more sustainable, or possibly unnecessary, in the future.

5.22   The Future of Health Information Exchange

With the dramatic growth in open APIs, FHIR adoption and the concept of health care app platforms, it is tempting to speculate about the future of health information exchange. Will specialized networks for health data sharing even be needed in an open standards-based, API-powered interoperable health informatics landscape? Many feel that cloud-based tools for coordinated, collaborative care would 'wrap' the complex mix of underlying EHRs and other systems into something far more useful and usable. These have now started to appear and some are coming from large companies with the expertise and resources to take on a hard, long-term problem. Figure 5.18 is a screen shot from the Salesforce.com Health Cloud that the company announced in late 2015. It represents a patient's external (professional) healthcare 'team' on the right and allows them to interact with each other, the patient and with their 'internal' team, on the left. This internal team might, as shown, consist of family members or even professionals the *patient* has chosen to consult. Unsurprisingly it is based on the company's robust CRM platform and suggests that new entrants will bring fresh technologies to healthcare.

InterSystems is a major supplier of core technology infrastructure (the Epic hospital information system uses the company's Caché® database technology). At the December, 2017 HL7 FHIR Applications Roundtable meeting the company showed that it is supporting FHIR as a data source as part of its HealthShare HIE technology (Fig. 5.19). It is also enabling SMART on FHIR apps to run against any data avail-

---

[41] http://www.rwjf.org/content/dam/farm/reports/reports/2015/rwjf423440

Fig. 5.18 Salesforce has developed a web-based platform for coordination and communication among patients' providers and family members. (Courtesy Salesforce)

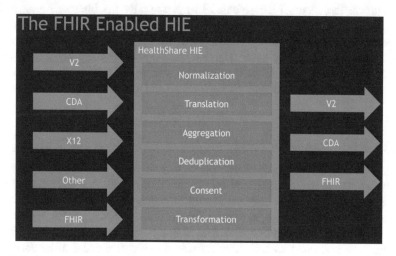

Fig. 5.19 A diagram of the InterSystems HealthShare HIE architecture as a FHIR-enabled informatics ecosystem. The HIE concept is still present, but it has also become a platform for FHIR apps. (Courtesy InterSystems)

able to the HIE (with proper identity and permissions). A video of the presentation is available and includes a sample of a SMART on FHIR app running against the aggregated data.[42]

---

[42] https://www.youtube.com/watch?v=QnSXxTgcGJI

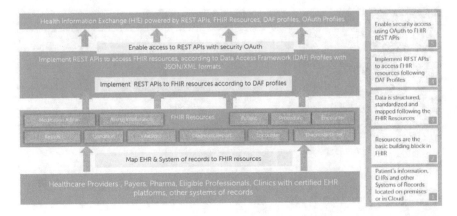

Fig. 5.20 A hypothetical architecture for a totally FHIR based HIE that could possibly also include some data transformation in the future to create a degree of semantic interoperability. (Courtesy Dr. Sandesh Prabhu, Youssef Serghat and NTT Data Services)

Figure 5.20 presents a hypothetical (at least at present) HIE architecture that is entirely FHIR-based. If such an environment existed and all the underlying EHRs and clinical tools were FHIR-enabled then essentially the Internet becomes the HIE. This would have seemed an implausible scenario a few years ago but today it seems to be virtually inevitable. Note in particular the inclusion of data transformation to presumably achieve some degree of semantic interoperability. REST based architecture is starting to find its way into HIE commercial products so this may not be a hypothetical idea for long.

5.23   Final Thoughts

These interesting and exciting new models for HIE in both the payer and clinical spaces suggest an architecture that should be much easier and less expensive to deploy and maintain. They also suggest that the rapidly growing FHIR app ecosystem may find a home beyond commercial EHRs. For example, it is entirely plausible that an architecture such as InterSystems' HealthShare could make data from non FHIR complaint sources available to FHIR apps that could work on individual or population level data. UHIN's activity in the payer space points toward the use of FHIR to provide more facile access to clinical data for use in the payer domain that we will explore in the next chapter. Given these advantages and existing efforts it seems inevitable that some REST based architecture will become the new standard for HIE. Most likely that standard will be FHIR. No matter the specific details, healthcare seems to be heading rapidly toward a more facile interoperable framework with the potential to address the long-standing need for solutions to the problems we discussed at the outset of this book.

Chapter 6
Payer Applications of FHIR

6.1 Interoperability in the Payer Space

Earlier we discussed how the Utah Health Information Network (UHIN) began as a secure, standards-based approach to exchange health insurance claims information. Today, the interaction between health care providers and payers is often retrospective when it could be done in real time. It is often episodic and it could be a far more efficient process if provider and payer information systems were able to share data electronically. This is yet another opportunity for interoperability and payers have begun to appreciate and act on the potential of FHIR to transform their business processes, particularly as they connect with providers.

In this chapter we discuss a domain that is effectively a "new frontier" for health informatics. We do this using three illustrative case studies – The Da Vinci Project, Humana's use of FHIR to help facilitate care and TIBCO's software platform – to illustrate how payers are starting to use FHIR to access clinical data in order to make more informed and timely decisions while reducing administrative burdens on providers with the goal of improving care quality, reducing costs and improving clinical outcomes.

6.2 The Da Vinci Project

The Da Vinci Project, convened under HL7, is an interesting forward-facing example of the use of FHIR outside of direct patient care. It is a private-sector led initiative "to establish a rapid multi-stakeholder process for addressing value-based care use cases that could be implemented on a national basis." The founding members of Da Vinci include payers, providers, HIT vendors, HL7 and federal agencies who are committed to making value-based care a reality. Toward that end, the overall goal of

© Springer International Publishing AG, part of Springer Nature 2018
M. L. Braunstein, *Health Informatics on FHIR: How HL7's New API
is Transforming Healthcare*, https://doi.org/10.1007/978-3-319-93414-3_6

the Da Vinci Project is to help payers and providers to positively impact clinical outcomes, quality, cost, and care management outcomes.

To promote interoperability across value-based care stakeholders, Da Vinci feels that the industry needs common: (1) standards, including HL7® FHIR®, (2) implementation guides, and (3) reference implementations to guide the rapid development and deployment of interoperable solutions on a national scale. In keeping with that, the initial Da Vinci efforts will include (1) identifying specific value-based care use cases that are amenable to national solutions using FHIR APIs and (2) rapidly developing implementation guides and reference implementations, and (3) field testing them to validate their readiness, effectiveness and efficacy."[1]

As of this writing, work is just getting underway as the cross-function group develops clear goals, operating guidelines and the necessary underpinnings for a loosely coupled project like this to function. The project aims to produce the initial two public use cases in 2018, and to have begun to evaluate a third use case. Of the countless potential use cases, the Da Vinci founders have narrowed the initial focus to the nine shown in Fig. 6.1.

The first two to be worked on are common, but still rich in complexity: 30 Day Medication Reconciliation and Coverage Requirements Discovery.

30 Day Medication Reconciliation Earlier we discussed transitions of care as a common source of medical error. A specific source of that error is a failure to have a complete view of a patient's medications. Automated medication reconciliation is the process by which a complete view is assured by reconciling the patient's medication records from the venues from which they are receiving care. Of course, it is also important to compare the record with what the patient is actually taking but that is not typically an automated process. The importance of medication reconciliation is demonstrated by its frequent inclusion in the Healthcare Effectiveness Data and Information Set (HEDIS)[2] used by health plans to measure care quality and in commercial health insurance contract requirements.

The 30 Day Medication Reconciliation use case focuses on a care team's ability to ensure a newly discharged patient and their providers have a clear view of the up-to-date medications a patient is currently taking. The need to identify early any confusion, problems or abandonment is tied to reduction in adverse outcomes for patients leaving acute care setting.[3] While the ideal state would be to have a shareable, curated list of validated medications for a patient, the initial focus for the use case is to capture the "attestation" from a care team member that the task is complete. Work is underway to determine the method, resources and actors in a FHIR based data exchange between Payer and Provider. It is envisioned that there are potential intermediaries acting on behalf of each participant including the EHR, population health vendor, or even an application being monitored by clinical pharmacist inside an integrated delivery network.

---

[1] http://www.hl7.org/about/davinci/index.cfm

[2] http://www.ncqa.org/hedis-quality-measurement

[3] http://www.ajpb.com/journals/ajpb/2014/ajpb_marapr2014/importance-of-medication-reconciliation-in-the-continuum-of-care

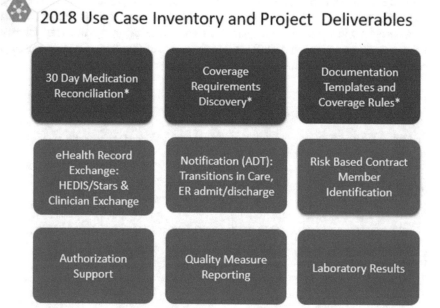

Fig. 6.1 The initial set of nine Da Vinci use cases. Of these, 30 Day Medication Reconciliation and Coverage Requirements Discovery are in initial development with a completion goal by the end of 2018. The other 2018 goal is to begin evaluation of the Documentation Templates and Coverage Rules. (Courtesy HL7 Da Vinci)

Coverage Requirements Discovery Understanding the specifics of each patient's insurance coverage and the requirements to obtain payment is a critical element for the transformation of care delivery. In value based care it is essential to expand a provider care team's access beyond just the appropriate clinical data about the patient prior to and during patient visits. Increasingly complex health plan design requires understanding of the benefits and services at the individual patient level at each encounter. The ability for a provider or supporting team member to understand the necessary pre-work activities and documentation required for a particular service or procedure will reduce the administrative burden on the care provider, and ensure that data and documentation are collected and defined while a patient is in the office instead of the often retrospective, start/stop nature of the current claims and prior authorization process. In an ideal workflow the provider would have a clear view of the requirements and documentation required to provide the appropriate services and procedures to the patient in their care in order to ensure a patient can get from diagnosis to care plan to the delivery of the recommended services.

Building each of the identified Da Vinci use cases will require a period of discovery, definition and incremental development. This will in turn drive the process of maturing from an initial to the eventual ideal data set required for value based care data exchange. The goal for the initiative is to leverage the knowledge and experience of the industry to deploy meaningful, usable how to guides and validate their approach. Where the proposed implementation guides merit standardization, the Da Vinci project will work with the appropriate HL7 working groups and leadership to bring recommendations and findings into the existing standards development process for eventual balloting.

Case Study: Humana

Humana Inc. is one of the largest profit, for-profit US health insurance companies with 14 million subscribers and some $53.8 billion of revenue. Patrick Murta, Principal Solution Architect, says the company is "committed to helping our millions of medical and specialty members achieve their best health." The company says that its goal is what it describes as a new kind of integrated care and it is focused on using FHIR to make that happen. For many years, Humana has recognized the importance of technology as a means toward this goal.

The company views its technology efforts as having occurred in three phases. The first administration phase, introduced web browsers and web services (before they were officially named web services) to enable real time eligibility and benefit checks, instant pre-authorization, real-time claims submission and decisions, payment remittance, and other real time administrative transactions.

The second phase began in 2011–2012, and it involved sharing clinical information with providers and subscribers. The company had finished a several year project to build a sophisticated analytics engine and statistical models (using terabytes of member data collected through the previous phase). This was able to provide insights into a member's status (care alerts, gaps in care, problem lists, medication issues, etc.) but the company needed a way to share this information with clinicians in a way that fit into their workflow. The company used data available to it to create a CDA formatted clinical summary document and made it available as a web page. To move ahead the company felt it needed to receive post encounter clinical information from providers. To do this it employed HL7 messages (V2 and V3), connections to HIEs, receipt of CCDs and some proprietary web APIs.

However, Humana's ultimate goal was to connect with providers and their staff as close to an integrated part of their workflow and it feels that "the adoption of FHIR has the potential to enable the truly integrated care delivery model that we aspire to while at the same time reducing costs and inherent inefficiencies in the system".

Mr. Murta goes on to say that "We are at an inflection point in our industry's technology. With clear movement and demand from stakeholders including patients, providers, payers, EHR vendors, and integrators, and clear direction and support from CMS, ONC, the House and Senate, and the White

House that APIs will be used to move integration forward, this must happen. We have a technology that, if nurtured, supported, and adopted, will make this happen."

Humana had years of classic service oriented architecture experience and was starting to explore RESTful open API JSON services when FHIR started gaining significant attention in 2014. After a learning period Humana says that the 'think FHIR first' approach has become more prevalent. It follows a six step FHIR development process:

1. Starting with an integration use case, a solution architect will evaluate the need and consider FHIR as a preferred model. Once recognized, this will typically be communicated back to the team in the form of an 'approach on a page' mentioning FHIR as the strategy with a brief explanation of the approach and benefits.
2. The solution architecture will identify where FHIR integration is appropriate, recommend the FHIR resource, and benefits of using FHIR for the use case. Use of an existing FHIR service is preferred since Humana is using a build to the standard once and reuse model of FHIR development.
3. Stories illustrating the use case and the role of FHIR are shared across the development and IT teams, to reinforce from both a business and technical standpoint why FHIR is being employed.
4. A virtualized version of the service with synthetic data is made available in a 'sandbox' to allow developers to test and validate. Since the FHIR resources are predefined, the interface remains fairly static throughout development and teams don't have to wait for completed development before testing interfaces.
5. QA testing is handled by a testing teams specializing in data and traceability to source. Performance and scalability testing is handled by another team.
6. Since much of the data processed via FHIR is sourced from fully functional but non-FHIR based systems, the architecture and eventual deployment will provide outward facing FHIR endpoints as shown in the top three layers of Fig. 6.2. These provide the external endpoint and management layer, the internal endpoint and management later, and the underlying services and data.

Humana is exploring a number of initial use cases that involve FHIR. In what follows, we will explore a few of them and consider some future possibilities.

Medication Based Care Alerts: This clinical decision support tool in invoked as prescribing clinicians are considering medication changes (Fig. 6.3) to provide insights about potential drug to drug, drug to disease, drug to food, or medication adherence. Humana feels that it is in a position to add value because of its analytic activities that may have derived insights from data collected from other providers or labs that are not available in the local EHR or e-prescribing system.

(continued)

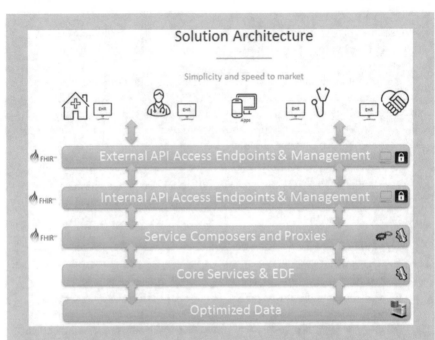

Fig. 6.2 This diagram of Humana's technology stack illustrates the several key roles that the FHIR API is playing and will play into the future. (Courtesy Humana)

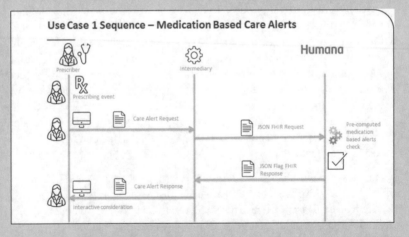

Fig. 6.3 This is a schematic representation of the interaction between a medication prescriber and Humana to get care alerts during the prescribing process in the patient encounter. (Courtesy Humana)

(continued)

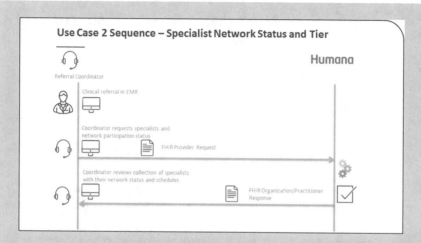

Fig. 6.4 A diagrammatic representation of a more facile and efficient provider and payer interaction for optimal specialist referral options. (Courtesy Humana)

Specialist Network Status Tier: Before a health insurance company can agree to pay for care by a specialist it must determine if that the provider is part of (is participating) in its care network. This process and scheduling of the referral visit, if approved, has historically involved time consuming manual processes. The revised process shown in Fig. 6.4 utilizes a FHIR API to electronically provide network status (participating/non-participating) and other provider information including schedule availability within the referring provider's workflow.

Post Encounter Clinical Document Submission: In particular circumstances (such as admission to the hospital) certain information must be provided to the payer after an encounter. The FHIR API illustrated in Fig. 6.5 provides a simple RESTful interface to publish real-time messages such as an HL7 V2/3 ADT message, a C-CDA document of even an image.

Payer Based C-CDA: As we discussed earlier, specialists often see patients who have been referred to them with little or no prior clinical information. This API provides an on demand clinical summary, within their workflow, to treating physicians who may not have any other supporting clinical documentation at the time of encounter. Another use for this service is telemedicine clinicians who do not have access to clinical documentation seeing a remote patient virtually.

Future plans for FHIR API development at Humana include:

An API to find appropriate, participating specialists for a patient in a given area.
An API enabling enhanced medication check that evaluates potential drug to drug contra-indications before prescribing.
An API which enables providers to automatically and electronically submit clinical document in support of a prior authorization request.

(continued)

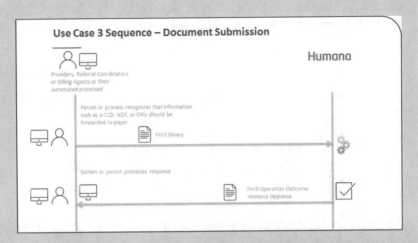

Fig. 6.5 A diagrammatic representation of post-encounter data sharing between the provider's office and Humana to streamline backend claims and clinical processing. (Courtesy Humana)

An API enabling an EHR or other clinical system to determine if prior authorization is needed for a specific clinical referral. If prior authorization is required, the EHR or app can automatically submit the required administrative information without requiring the referral coordinate to log into another application.

An API that uses real-time analytics on an eligibility and benefits request to determine the likelihood that an ER visit is underway, or has occurred, and pro-actively notifies the PCP of the ER visit.

An API which enables medical and consumer devices to submit information in real-time for pro-active issue detection and care team coordination.

An API for better integration of Humana's internal care management clinicians (and associated care plans) with external care team members and apps.

Through its population health technology company, Transcend Insights, the company is building a collaboration and population health management core to bring patients, providers, and caregivers together for mutual benefit. FHIR and advanced analytics based on FHIR document objects are key foundational technologies. The company says this model is the key underpinning of the company's integrated care delivery model and it is currently being tested community wide in a Kentucky town.

Later we will discuss the new CMS Blue Button 2.0 API that provides Medicare beneficiaries with access to 4 years of their claims data. Humana plans to use this information to enhance their predictive modeling, identification of gap closure, provision of care alerts and more. This will be the company's first use case to engage with the member community using member facing FHIR apps.

(continued)

Case Study: TIBCO Software

TIBCO is a Palo Alto based private company (at the time of this writing) founded in 1997 that provides integration, analytics and event-processing software for over 10,000 customers to use in on-premises or cloud computing environments. The company has a diverse set of healthcare clients ranging from delivery systems, such as universities; charitable, non-profit medical providers; EHR vendors; and a number of payers, third party administrators and service providers.

The TIBCO Connected Intelligence Platform helps digitize the way health insurance payers operate by incorporating multiple digital channels. Potential benefits of this capability include creating a seamless and secure connection to any external FHIR API to improve the way payers obtain electronic health records, and utilizing the data to improve their business processes and enrich their product offerings. Examples include tools and the related information to more easily make informed decisions on risk profiling, claim approvals, fraud detection, and other factors.

While HIPAA (the Health Insurance Portability and Accountability Act) mandates an EDI formatted protocol for exchanging eligibility, claim, prior authorization and payment data between providers and payers. The primary example of this is the HIPAA 837 claim that is typically used by providers and hospitals to file for payment.[4] There is currently no mandated standard for providers, hospitals and payers to exchange electronic health records. These records are necessary to further determine whether or not a procedure or service is authorized, a claim should be paid, a risk score is accurate, or other uses. Obtaining electronic health records is an ideal use case for FHIR.

TIBCO presented a demonstration of this at the December, 2017 HL7 FHIR Application Roundtable and Payer Summit. The solution enables a payer to obtain specific health records as FHIR resources via a manual trigger or an automatic trigger.

In a manual trigger someone from the payer side initiates the request, for example, in the case of risk adjustment. A health insurance company that offers Medicare Advantage (MA) plans (an alternative to traditional Medicare that typically offers expanded coverage within a provider network) can receive more accurate payments from CMS (Center for Medicare and Medicaid Services) for the healthcare expenditures of its members based on the health status of those members. If the health status of a patient is different than originally predicted, the insurance company will receive an adjusted payment from CMS. After a patient's file is reviewed and it is determined that additional information is needed to substantiate a higher-risk diagnosis, the payer can utilize the TIBCO solution to request the patient records.

Figure 6.6 is a sample FHIR user interface in which a Patient ID (4342010) is entered, initiating a workflow process in TIBCO BusinessWorks™ software. The example shown in Fig. 6.7 makes an API call to the Cerner FHIR

(continued)

---

[4] https://www.cms.gov/Outreach-and-Education/Medicare-Learning-Network-MLN/MLNProducts/Downloads/837P-CMS-1500.pdf

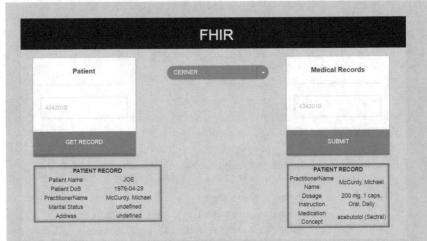

Fig. 6.6 Sample FHIR user interface utilized by either a payer or provider to extract required patient information from a FHIR enabled EHR. (Used with the permission of TIBCO Software Inc. © TIBCO Software Inc. All rights reserved.)

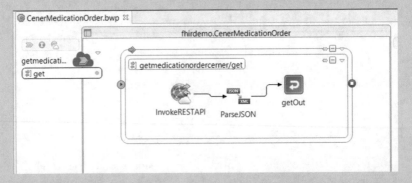

Fig. 6.7 TIBCO's BusinessWorks software automates the process of making a REST API call to collect to the Cerner FHIR server to retrieve the required patient information, in this example a medication order. (Used with the permission of TIBCO Software Inc. © TIBCO Software Inc. All rights reserved.)

server to retrieve the patient information or medical record. The retrieved patient data are displayed in the lower half of the UI in Fig. 6.6.

An automatic trigger occurs when technology determines that additional information is needed to pay a claim or if a patient is at high risk, based on values contained in a data file. For example, a particular diagnosis code might always require additional information. As shown in Fig. 6.8, TIBCO BusinessWorks software detects this in the HIPAA 837 claim data and auto-

(continued)

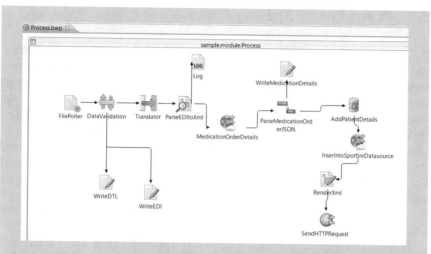

Fig. 6.8 TIBCO BusinessWorks software automates a FHIR API call process to retrieve the required medical documents in order to authorize a claim. (Used with the permission of TIBCO Software Inc. © TIBCO Software Inc. All rights reserved.)

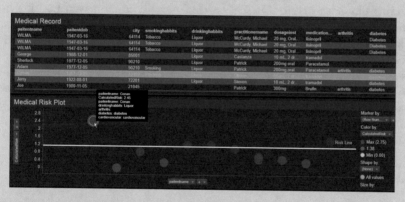

Fig. 6.9 TIBCO Spotfire® software can be used in the visual exploration of the appropriate premium to bill a customer given the risk that they pose to the payer. (Used with the permission of TIBCO Software Inc. © TIBCO Software Inc. All rights reserved.)

matically activates a FHIR API call to retrieve the necessary medical records from the provider noted in the claim.

Medical record data is analyzed to identify patients that are considered high risk and to assign a risk score to each patient. As shown in Fig. 6.9, the risk scores are then viewed in a dashboard report in TIBCO Spotfire® software highlighting which patients fall above, below or at the median risk level. Additional medical record data may be requested using the FHIR API for high risk patients.

6.3   Future Uses of FHIR by Payers

It is worth considering possible future uses of FHIR by payers. One potential example are enhancements to one of the UHIN services we discussed earlier. You may recall that, using UHIN, payers can make a FHIR query to request and receive needed clinical information to process claims or prior authorizations. When physicians propose unusual or expensive tests, medications or procedures for one of their patients, the payer may need clinical justification before agreeing to pay for it. Currently this prior authorization process can be largely manual at the provider and payer end. Medical office staff may copy information from the patient's chart and fax it to the payer where trained nurses may prepare an abstract and input it into the payer's system. If authorization requires further review it is done by a physician. It is possible that this cycle has to be repeated until all the needed information is obtained.

Here again the granular nature of FHIR could provide a more facile solution. It is conceivable that a provider facing FHIR app could understand what clinical information is needed to authorize the specific test, medication or procedure being ordered and could automatically abstract it from the patient's chart for review by the ordering physician before it is sent on electronically to the payer. Looking even further out, provider organizations that are more actively managing patients in value-based contracts could perform an initial review of the clinical criteria against a FHIR service that exposes clinical criteria supplied by their payer partner and instantaneously "approve" the test, medication or procedure if sufficient clinical justification exists. This use case is a particularly interesting and very practical example of how FHIR could expedite the use of clinical data for an important secondary purpose.

6.4   Recap

This chapter concludes Part II of the book. The rapidly growing interest in FHIR in the payer community illustrates the potential of FHIR to help address interoperability challenges well beyond the traditional delivery of care in a clinical setting. The case studies presented in this chapter provider examples of how data made more available using FHIR can bridge a historic chasm separating the provider and payer spaces with the promise of better care at lower cost and substantial simplification of what are now complex administrative workflows and processes.

Part III
Interoperability Essentials

Chapter 7
Data and Interoperability Standards

7.1 Introduction

Data and interoperability standards are the virtually ubiquitous plumbing that underlie most contemporary health informatics systems and tools. We've already gotten a glimpse of that when we encountered LOINC codes in the lab test HL7 V2 message.

Given the complexity of healthcare, it should not be surprising that this is a complicated topic, so we will divide it into three chapters. In this initial chapter, we will first discuss the evolution of standards and then we will focus on standards for representing health data. In the next chapter, we will discuss how that data is packaged, transported and shared using interoperability standards developed before FHIR. In the third chapter, we will cover FHIR in some detail.

7.2 Why Standards?

Why do we need data and interoperability standards? We have already discussed that as part of examining semantic interoperability, but it is useful to consider it more specifically here. First some definitions. The term syntax refers to grammatical structure whereas the term semantics refers to the meaning of the vocabulary symbols arranged with that structure.

Figure 7.1 presents a simple illustration from Grant Wood at Intermountain Healthcare that helps to make the differences and ambiguities in representing healthcare data clear using a seemingly simple data element, the patient's gender. System A represents males with a "1" and females with a "0". That representation is reversed in System B. The two systems are using the same 'language' of 1's and 0's but the 'words' have different meanings so they cannot interoperate without some intermediate translation process that make them semantically interoperable by

© Springer International Publishing AG, part of Springer Nature 2018
M. L. Braunstein, *Health Informatics on FHIR: How HL7's New API is Transforming Healthcare*, https://doi.org/10.1007/978-3-319-93414-3_7

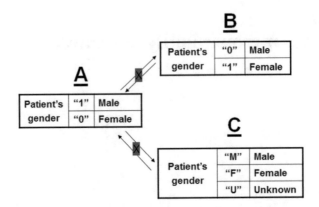

Fig. 7.1 An illustration of both syntactic and semantic differences in the representation of a seemingly simple data item, a patient's gender. Systems A and B use the same 'language' of 1's and 0's but reverse their meaning. System C uses a different language and introduces and a new 'word' or value. Usefully sharing or combining data among these systems requires some mapping or conversion to a common standard to bridge these differences. (Courtesy Grant Wood, Intermountain Healthcare)

mapping one to the other or both to some common representation. Should you mix data from these two systems without such an intervening "curation process", gender might be impossible to determine accurately for care delivery, reporting and other purposes and patient care issues could even be compromised.

In System C, "M" is used for male and "F" for female. System C, roughly speaking, has a different syntax from systems A and B – it uses a different "language" to represent gender. Moreover, System C recognizes that gender may be ambiguous and represents that with a U, a concept that the other two systems do not deal with. Interoperability between System C and the other two systems would require translation from its 'language' to theirs or translation of all these approaches to a common form – a standard. Since undetermined gender is not represented in systems A and B, some accommodation for that would also have to be made, particularly since undetermined gender could be of clinical importance and its rate could be of interest for research or reporting purposes.

Interestingly this exact situation arises in FHIR as compared to the earlier HL7 Reference Information Model (RIM) that was the basis for the Clinical Document Architecture (CDA) we discussed earlier. Many fields in FHIR are constrained to a specified list of values, a 'value set'.[1] The FHIR value set for a patient's gender contains four entries male, female, other and unknown. The earlier value set for the HL7 Reference Information Model has the three gender values: male, female and undifferentiated.[2]

---

[1] https://www.hl7.org/fhir/codesystem-administrative-gender.html

[2] http://www.hl7.org/documentcenter/public_temp_60E0324C-1C23-BA17-0CD7909FBEC-0CE5B/standards/vocabulary/vocabulary_tables/infrastructure/vocabulary/AdministrativeGender.html

If something as seemingly simple as gender can lead to this much complexity, imagine what happens with the concept of a patient's diagnosis which is inherently somewhat subjective, can have thousands of possible values and, as we will see, can be described at varying levels of detail! We got a sense of that in the discussion of semantic interoperability earlier. That is why data standards have been developed, but, as we will also soon see, they can themselves get quite complex and can be problematic as a result. In fact, as we shall see, this is far from a hypothetical issue and it has plagued many of HL7's standards efforts introduced between the successful messaging standard and the FHIR development effort which is trying hard to limit complexity.

7.3 Standards Structure and Purpose Evolution

Standards have evolved over many years in an effort to encompass more aspects of medicine; to code them in more detail; and to adapt as technology changes. We will divide this evolution into three dimensions: structure, purpose, and technology. We will discuss the first two in this section and technology evolution in the next one.

The **structure** of standards has evolved in large part to take advantage of the capabilities of computing. Early data standards were lists, such as medical diagnoses, laboratory tests, or medications. We will refer to these list standards as a Classification. As the use of computers in healthcare grew, so did interest in the standards community to describe more detail and to represent relationships among clinical concepts. We will refer to a standard that can code for relationships as an Ontology. The goal here is often semantic interoperability.

If you viewed it, you may remember the graph of clinical relationships in the Applicadia video shown in Fig. 7.2 and the nuanced, multi-layered representation of clinical details, concepts and relationships it portrays.[3] The graphs shows that Applicadia is using many coding systems to represent concepts and their characteristics. These include ICD-10 and SNOMED-CT which are both complex hierarchical ontologies that are able to represent the concept relationships that are at the core of what Applicadia is trying to do.

The **purpose** of standards has also evolved. Pre-computing, all of the early standards were for data. Physicians would use the International Classification of Diseases (ICD) in their charting, primarily for billing. The clinical laboratory would use the LOINC classification we touched on earlier when we looked at HL7 messages for lab tests, and the pharmacy would use one of several relatively simple classifications to represent the medications they dispensed.

In the 1980s computers were increasingly installed by hospitals. The focus was largely on the revenue producing departments, such as billing, laboratory, pharmacy and radiology. The lab and the nurses' stations needed to exchange orders and test results. Everything done to patients that was billable needed to flow from the nurses'

---

[3] https://www.youtube.com/watch?v=INWxXJThIZE

Fig. 7.2 A graph of clinical relationships from the Applicadia video illustrates a nuanced, multi-layered representation of clinical details, concepts and relationships using many data standards including the ICD-10 and SNOMED-CT ontologies. (Courtesy Applicadia)

station to billing. Medication orders needed to go from the nurses' station to the pharmacy and the record of their administration might, in some cases, be shared. All of these entities needed to know when a new patient was admitted including their usual demographic information, medical record number and location in the hospital.

However, often each of these departments of the hospital was using a specialized and independently developed, non-interoperable, proprietary software module. This approach was called 'best of breed' and typically each department procured the system it felt best met its needs. This was actually the only feasible approach at the time since more complete, integrated 'whole hospital information systems' were not yet very widely available and those that did exist (Shared Medical Systems and MCAUTO were the main vendors) were based on expensive main frame computer technology and did not provide very robust integration even within their own system.

After some awkward intermediate 'solutions' (including connecting terminals from the nursing unit to each of the systems so that a user could use all of the systems). Solutions emerged based on the relatively new concept of local area networks. One in particular was StatLAN created by Simborg Systems Corporation. The company's Founder and Chief Executive Officer, Don Simborg, MD, was a cofounder of HL7. The history of the evolution of these technologies along with a history of HL7 itself is posted by the independent Dutch health standards consulting company, Ringholm.[4]

Ringholm also discusses the history of the HL7 messaging standards that preceded V2. "The HL7 protocol dates back to the late 1970s when its precursor was

---

[4] http://www.ringholm.com/docs/the_early_history_of_health_level_7_HL7.htm

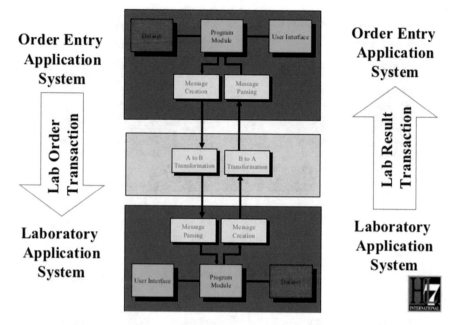

Fig. 7.3 Lab test order and result flow using HL7 V2 messaging. Note in the middle that data translation and message parsing are required to make this happen. (Courtesy of HL7)

developed at University of California at San Francisco (UCSF) Medical Center and first implemented in production in 1981. HL7 v1 and V2 are essentially refinements of the UCSF protocol. X12 and ASTM E1238 have had a large impact on the development of HL7. The HL7 organization matured from a small ad hoc working group in 1987 into a full blown standards development organization within its first 5 years."

As shown in Fig. 7.3, using the HL7 V2 messaging technology, a physician could order a lab test or medication at the nurses' station and the order to perform that test could go electronically to the clinical laboratory and the result could come back to the nursing station using a different message from the same standard. The technical flow to accomplish this is shown in the middle of the figure.

The next evolution of the purpose and structure was standards for clinical documents. A message would typically be an order for a single lab test or medication. However, the physician who would care for a patient after hospital discharge required a complete summary of that patient's care. This is the role of document standards. You worked with one of them – the HL7 CCD – if you did the HealthVault exercise. As is often the case with standards, more than one was developed by different organizations (in this case for essentially the same purpose), as shown in Fig. 7.4. The Continuity of care Record (CCR) on the right was developed by a group that included the ASTM International, a standards organization, and HIMSS, the huge health information technology industry group we mentioned earlier. The Continuity of Care Document (CCD) derives from HL7's Reference Information

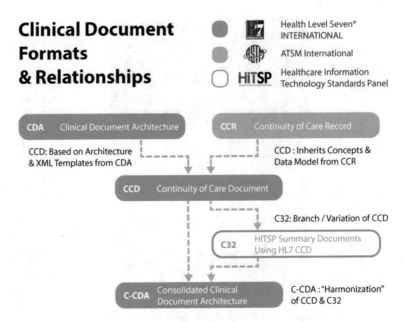

Fig. 7.4 Often, multiple organizations create different standards for the same thing. In this case it's an XML formatted electronic clinical document for use in transitions of care such as a patient discharge from the hospital or a referral from a PCP to a specialist. This diagram shows the progression of separate HL7 CDA, ATSM CCR and HITSP C32 standards for a patient summary to the current unified C-CDA based version of the CCD. (Courtesy HL7)

Model (RIM). Both are formatted in XML. As you can see in the figure they were reconciled into the Continuity of Care Document you may have worked with in the HealthVault exercise. You will note a branch in the diagram to the HITSP C32 which is a constrained version of the CCD document defined by the Health Information Technology Standards Panel (HITSP) for information exchange. When we discussed CIMI we mentioned FHIR profiles and the idea of specifying the specific content and values allowed within a standard for a specific use case. This is another example of that. The Consolidated Clinical Document Architecture or C-CDA was designed to harmonize the original CCD and the C32 in order to create a single standard. However, if you work in health informatics, you may still run into a CCR, CDA or C32 formatted clinical document.

More recently, as computers have become much more powerful, standards have been evolving to represent clinical processes and workflows. We will not cover these standards in detail but this is an extremely important area for future work and is recognized as such by the FHIR standards development effort.

We have previously mentioned Integrating the Healthcare Enterprise (IHE), an initiative by healthcare professionals and industry to improve the way computer systems in healthcare share information. It says that its goal is to promote the coordinated use of established standards to address specific clinical needs in support of optimal patient care because information can be passed seamlessly from system to system within and across departments and made readily available at the point of care.

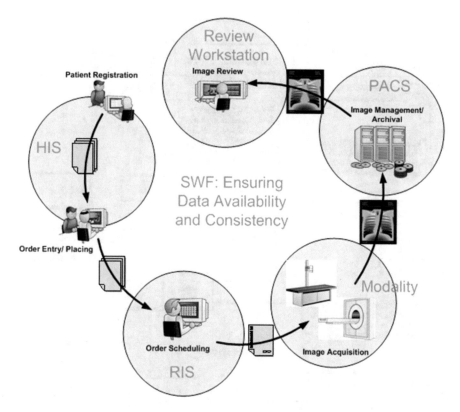

Fig. 7.5 A high level illustration from IHE of the Scheduled Workflow (SWF) that integrates the ordering, scheduling, imaging acquisition, storage and viewing activities associated with radiology exams. (Courtesy IHE)

Figure 7.5 is a high level illustration from IHE of the Scheduled Workflow (SWF) that integrates the ordering, scheduling, imaging acquisition, storage and viewing activities associated with radiology exams. This would involve established standards from HL7 as well as the widely used Digital Imaging and Communications in Medicine (DICOM) standard for the transmittal, storage, retrieval, printing, processing, and display of medical images.[5] Analysis of this workflow leads to identification of the information systems or applications that produce, manage or act on in-formation (actors) and the work that is performed between them (transactions) as shown in Fig. 7.6. This is the basis for IHE Profiles that describe clinical information and workflow needs and specify the actors and transactions required to address them. The reference is to the specific IHE Profile for the ordering, scheduling, imaging acquisition, storage and viewing activities associated with radiology exams. Radiology was the initial use case for IHE and is probably the area where its profiles are most widely used.[6] We will discuss vRad Radiology using DICOM and analytics to improve workflow in the final chapter.

---

[5] https://www.dicomstandard.org/

[6] http://wiki.ihe.net/index.php/Scheduled_Workflow

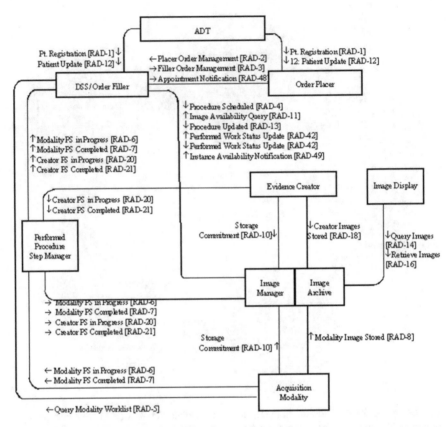

Fig. 7.6 More detailed analysis based on the overview creates a diagrammatic view of the information systems or applications that produce, manage or act on information (actors) and the work that is performed between them (transactions). This is the basis for IHE Profiles that describe clinical information and workflow needs and specify the actors and transactions required to address them. (Courtesy IHE)

Properly implemented, IHE standards could provide benefits including greater efficiency and a reduction in errors caused by unnecessary variations in patient care. Implementation does involve many areas of the hospital and systems so it requires a concerted effort. Nevertheless, given the cost of modern imaging equipment, this is a good use case and the IHE Radiology group says that hundreds of commercial radiology-related information systems have incorporated its solutions globally.[7]

---

[7] https://www.ihe.net/Radiology/

Case Study: Juxly

It seems clear that FHIR could be the basis for many workflow and process oriented clinical and administrative tools. Juxly was founded by Howard Follis, MD, a practicing urologist "with a deep passion for usability in medical software." He says he started the company to provide discrete point-of-care apps that reduce complexity for providers and enhance their ability to provide the best possible patient care.[8]

The company presented an early example of this, its Timeline FHIR app, at the December 2017 HL7 FHIR Applications Roundtable meeting.[9] The Timeline displays a patient's entire medical record organized temporally. The goal is to make the EHR data more usable and efficient for providers while they are with the patient at the point-of-care. The key innovation is displaying the temporal relationships between the data to try to give individual data points context that providers often need when seeing patients.

For example, in the Timeline screenshot shown in Fig. 7.7, this patient's last outpatient visit was immediately preceded by a renal arteriogram, a new prescription written for Lantus insulin and administration of a flu vaccine. Ten

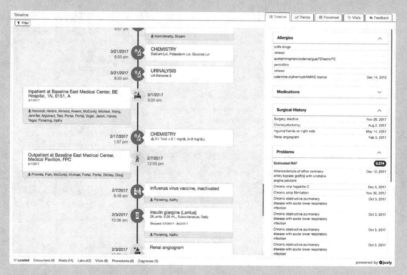

Fig. 7.7 Juxly's Timeline displays a patient's entire medical record organized temporally. The goal is to make the EHR data more usable and efficient for providers with the patient at the point-of-care. The key innovation is displaying the temporal relationships between the data to try to give individual data points context that providers often need when they are seeing a patient. (Courtesy Juxly)

(continued)

---

[8] https://juxly.com/our-story/

[9] https://www.youtube.com/watch?v=Ock9eohrnuo&feature=youtu.be

days after the visit, a Chemistry panel revealed an abnormal Bilirubin level. After the patient was admitted to the hospital on 3/1/2017, all of their labs, x-rays, & medications can be viewed in the context of the visit. All of the data displayed on the Timeline can be drilled down directly from it by clicking the icons in the midline to display more detail.

All data can be filtered & customized by the Filter button at the upper left and viewed via the flowsheet and vitals tabs at the upper right. The feedback tab accesses a Quick Start Guide or 6 short app training videos. Providers can also send feature requests and bug reports to the company via the feedback tab.

A brief demonstration video of the app integrated into the Allscripts™ Sunrise™ EHR is posted on the company's site. Early on it also illustrates how launching a FHIR app is built into the menu structure within that EHR.[10] A similar launch technique is used by Epic and Cerner, the two most widely installed EHRs in major health systems.

Juxly also provides an interesting Trends view of the chart as shown in Fig. 7.8. Physicians often complain that they have to navigate around in their EHR to find all data relevant to the current patient or problem. This view

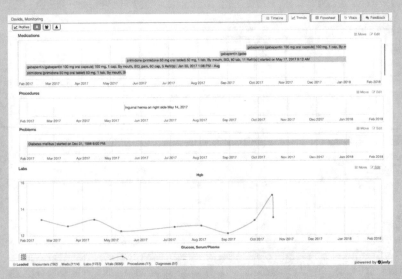

Fig. 7.8 Juxly's Trend view graphically brings together all the relevant clinical data for a chosen diagnosis over the same time frame making it easier for clinicians to see relationships among the various clinical data elements. An example might be the impact that a new drug or a change in its dosing (shown at the top) has on laboratory test results (shown further down). (Courtesy Juxly)

(continued)

---

[10] https://mp311339pt.cdn.mediaplatform.com/vod/allscripts.primetime.mediaplatform.com/0a51ed7c-953c-448a-845e-e73bf27acd47/video-1501003076692.mp4

graphically brings together all the relevant clinical data for a chosen diagnosis over the same time frame making it easier for clinicians to see relationships among the various clinical data elements. An example might be the impact that a new drug or a change in its dosing (shown at the top) has on physiologic measurements (shown further down).

Dr. Follis says they knew what findings to link to each diagnosis based on cases marked by individual providers as typical of their treatment of that particular diagnosis. This is, of course, somewhat reminiscent of how the Praxis EHR creates the note that a physician is likely to write based on their prior treatment of the same presenting complaint.

7.4 Standards Technology Evolution

The final dimension of standards evolution is **technology**. Early messaging standards used EDI/X12 – itself a standard that had evolved in other industries to automate business processes such as ordering, invoicing and payment. EDI/X12 dates from the early days of computing when memory and storage were dear so, as we saw earlier in the example of an HL7 V2 lab test message, it is quite cryptic and compact. With some effort, you could tell that the message we looked at involves a cell count and differential and reported specifically on the patients' hematocrit and erythrocyte count. The results and normal ranges were fairly easy to discern but most of the other details are not obvious without a guide.

Another example of EDI/X12 in healthcare is the virtually universal HIPAA 837 electronic claim we discussed earlier. You may recall that in the discussion of HIPAA you learned that a primary reason for the law was to move the industry to electronic financial and administrative transactions. The 837 electronic claim was a result and it is a particularly complex transaction that is also in the EDI/X12 format used by HL7 messages. Given its complexity, there are a number of commercial guides and courses devoted to properly entering the data. Combining a complex transaction like this one with the cryptic EDI/X12 format definitely creates challenges for those trying to create or interpret claims.

Newer messaging and document standards use XML, a more modern syntax that is verbose but has the advantages that it is more human readable and can be rendered in a browser. You saw that in the HealthVault exercise.

Figure 7.9 is a different lab test result from the one we looked at earlier when we discussed the HL7 V2 EDI/X12 message format. Here, it's in the newer HL7 V3 XML format. While it is still not particularly readable, we can more easily tell that it is a blood glucose level after 12 h of fasting and the value of 182 is high. We can also see that the normal ranges are for this test are 70–105.

```
<observationEvent>
    <id root="2.16.840.1.113883.19.1122.4" extension="1045813"
        assigningAuthorityName="GHH LAB Filler Orders"/>
    <code code="1554-5" codeSystemName="LN"
          codeSystem="2.16.840.1.113883.6.1"
          displayName="GLUCOSE^POST 12H CFST:MCNC:PT:SER/PLAS:QN"/>
    <statusCode code="completed"/>
    <effectiveTime value="200202150730"/>
    <priorityCode code="R"/>
    <confidentialityCode code="N"
        codeSystem="2.16.840.1.113883.5.25"/>
    <value xsi:type="PQ" value="182" unit="mg/dL"/>
    <interpretationCode code="H"/>
    <referenceRange>
        <interpretationRange>
            <value xsi:type="IVL_PQ">
            <low value="70" unit="mg/dL"/>
            <high value="105" unit="mg/dL"/>
            </value>
            <interpretationCode code="N"/>
        </interpretationRange>
    </referenceRange>
</referenceRange>
```

Fig. 7.9 A laboratory test result in the newer HL7 V3 XML based messaging format is easily for humans to read and illustrates that V3 is based on the HL7 Reference Information Model (RIM). (Courtesy HL7)

A more significant difference is that this V3 message is based on the HL7 Reference Information Model (RIM) that we mentioned earlier as the basis for the CCD and for some of the value sets used in FHIR. You may recall that the V2 lab test message referred to the LOINC code 18768-2 but there was no surrounding contextual information to make that clear. Here the LOINC code of 1554-5 almost gets lost in the surrounding contextual information that explicitly tells us that the 'codeSystemName' is 'LN' for LOINC and that its 'codeSystem' is '2.16.840.1.113883.6.1'. Search for that number and you'll easily verify that it links to LOINC. It's an object identifier or OID standardized by the International Telecommunications Union (ITU) and ISO/IEC for naming any object, concept, or "thing" with a globally unambiguous persistent name. Look the number up on oid-info.com and you'll find that LOINC codes are maintained by the Regenstrief Institute.

Below that is the LOINC 'displayName', itself a rather complicated five or six-part structure we will discuss later in this chapter. Once again, we've encountered an example of complexity introduced to provide more functionality and precision in a coding system. A simple LOINC code in V2 is now three lines of information in V3. It is all potentially useful information for some but, for many, it is just more to parse through to get what they need. Some would argue that, in totality, all of this added information and complexity has made V3 and RIM impractical in the real world. That view seems to have prevailed and it is clearly a key goal of the FHIR development effort to use the best parts of the RIM while keeping complexity under control.

7.5   The Key Data Standards

The six key data standards we will now discuss are:

- International Classification of Diseases (ICD-10)
- Current Procedural Terminology (CPT)
- Logical Observation Identifiers Names and Codes (LOINC)
- National Drug Code (NDC)
- RxNorm
- Systematized Nomenclature for Medicine (SNOMED-CT).

ICD and CPT are widely used in the US because, in most cases, they are required for medical billing. CPT and NDC are largely US specific. They are classifications, although CPT increasingly has sub-codes to provide more details about a procedure for more precise billing. LOINC is not quite an ontology but provides significant details about clinical tests and is in use internationally. ICD-10 and SNOMED-CT are internationally used increasingly complex ontologies capable of representing clinical relationships among their elements.

When we discuss FHIR later on, you will see that it builds on most of these data standards and FHIR resources specifically reference them.

7.6   International Classification of Diseases

The International Classification of Diseases (ICD) is the oldest data standard, dating directly back to the 1800s and more indirectly to earlier centuries when researchers became interested in the causes of human mortality. One of these was John Gaunt who, in the mid-1600s, published some of the first research on the causes of human mortality. The link provides an interesting history of ICD you can read if you desire.[11]

Traditionally ICD was a list or classification of medical diagnoses maintained by the World Health Organization (WHO) and updated every 10 years. ICD-10, the current version, was adopted in 1994, and work is well underway on ICD-11 for its scheduled release in 2018. The US finally adopted ICD-10 on October 1, 2015, well after most other countries.

The switch was a substantial effort because ICD-10 is a major quantitative and qualitative expansion. ICD-9 has 13,000 codes while ICD-10 has some 68,000 codes to represent very specific clinical details. Note in particular that ICD-10 can provide laterality so, Fig. 7.10 which is an example for breast cancer, illustrates that, unlike ICD-9, ICD-10 can indicate in which lower-outer quadrant the disease was located.

---

[11] http://www.who.int/classifications/icd/en/HistoryOfICD.pdf

ICD-9: 174.5	**ICD-10:** Three possible codes
Malignant neoplasm of female breast, left-outer quadrant	**C50.511:** Malignant neoplasm of **right** female breast, left-outer quadrant
	C50.512 Malignant neoplasm of **left** female breast, left-outer quadrant
	C50.519 Malignant neoplasm of **unspecified** female breast, left-outer quadrant

Fig. 7.10 One of the major differences between ICD9 and 10 is the over 50% expansion of the number of codes allowing, as shown here, for laterality to be represented. The ICD-9 code 174.5 maps to three ICD-10 codes depending on the laterality of this patient's disease

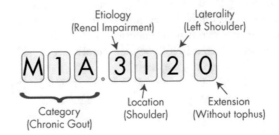

Fig. 7.11 The ICD-10 coding system provides important information about the cause, location and manifestations of gout most of which are absent in ICD-9. This results in 162 gout related codes in ICD-10 versus 13 in ICD-9. Many healthcare providers complain about the added complexity of using this new coding system but its benefits for more precise billing, patient care and clinical research should be clear. (Courtesy AAPC)

Beyond size, ICD-10 is an ontology capable of representing clinical relationships not represented in ICD-9. For example, as shown in Fig. 7.11, ICD-10 can encode the fact that a patient has gout affecting their left shoulder but that they have not yet developed a uric acid deposit, called a tophus, in that shoulder. ICD-9 cannot specifically code for gout located in the shoulder, much less the left shoulder. It can code for gout in an unspecified joint with or without a tophus or for a tophus in the ear (a common manifestation of gout which is presumably why it is specifically included as a code). Note also that the source of the gout can be coded, in this case as renal impairment, insufficient functioning of the kidneys which remove uric acid, the causal factor for gout if found in excess. ICD-10 can also code for lead or drugs as the cause of gout along with idiopathic gout of unknown origin. In all there are 162 ICD-10 codes for gout and its various causes, locations and manifestations. There are 13 ICD-9 codes whose description contains 'gout' or 'gouty'.

ICD codes are required on health care claims. Potential uses of the added detail in ICD-10 could include more appropriate provider reimbursement and preventing fraud caused by duplicate billing for the same service. However, as the Applicadia video demonstrates, the complexity of the coding system is itself a problem for both providers and payers. You will find ICD codes in FHIR Resources.

7.7   Current Procedural Terminology (CPT)

Current Procedural Terminology (CPT)[12] is a classification of medical procedures maintained and updated annually by the American Medical Association. Like ICD, it is required for virtually all reimbursement for healthcare services.

CPT codes divide into three categories:

Category I codes are five digit numbers for widely performed procedures and are divided into sections for anesthesiology, surgery, radiology, pathology/laboratory and medicine.

Category II codes are for the collection of quality and performance metrics and are four digits followed by an "F".

Category III codes are also four digits, followed by an "I", and are temporary, to allow for new or experimental procedures.

As you can see in Fig. 7.12, each code has a full, medium, and short description, as you can see here, by the various levels of detail for naming a flu vaccination for different purposes.

Given its use in billing, a CPT code may provide details necessary to determine the proper charge. In Fig. 7.13 CPT codes for a psychotherapy visit indicate the length of the visit and the amount of provider-patient interaction. The charge would be more or less depending on the code used.

Figure 7.14 presents the factors that determine the code and, hence the charge, for an initial new patient office visit. These include how extensive the history obtained by the provider from the patient was, the type of physical examination and the complexity of the medical decision. Clearly, there is some subjectivity to the selection of the code and the potential for a practice called 'upcoding' to increase reimbursement.

CPT codes may appear to be simple but, given their critical role in billing, and subtleties such as we have discussed, selection of the right code is important and

Description	
Full	Influenza virus vaccine, trivalent, derived from recombinant DNA (RIV3), hemagglutinin protein only (HA), protein and antibiotic free, for intramuscular use
Medium	Influenza Virus Vaccine, Trivalen RIV3 PRSR FR IM
Short	Flu Vacc RIV3 No Preserv

Fig. 7.12 CPT provides a full, medium and short description for each code. This can be useful in displaying CPT descriptions depending on the amount of space available, such as on a phone

---

[12] https://www.ama-assn.org/practice-management/cpt-current-procedural-terminology

CPT Psychology Code	Session Time	Minimum Face-to-Face Time
90833	30	16
90836	45	38
90838	60	5

Fig. 7.13 CPT codes for a psychotherapy visit include the length of the appointment and the amount of face time with the provider. Both can factor into the amount the provider will be reimbursed

	Office Visit Coding Guidelines: New Patient		
Code	History	Exam	Decision Making
99201	Problem Focused	Problem Focused	Straightforward
99202	Expanded Problem Focused	Expanded Problem Focused	Straightforward
99203	Detailed	Detailed	Low Complexity
99204	Comprehensive	Comprehensive	Moderate Complexity
99205	Comprehensive	Comprehensive	High Complexity

Fig. 7.14 The CPT code of a new patient office visit includes the amount of history that is determined, the type of physical examination and the complexity of the medical decision made by the provider during the visit. All of these factors can be subjective leading to the potential for 'upcoding' to increase reimbursement

billing personnel require training to code correctly. The goal typically is to submit the largest, but hopefully legitimate, bill.

7.8 Logical Observation Identifiers Names and Codes (LOINC)

Earlier, we mentioned the Regenstrief Institute in Indiana in our discussion of the Indiana Health Information Exchange. We also saw the use of LOINC codes in an HL7 laboratory test result message. The institute also developed and maintains the LOINC code system for laboratory tests and clinical observations.

Starting in 1994, health informatics pioneer, Dr. Clement "Clem" J. McDonald, organized the LOINC Committee to develop a common terminology for laboratory and clinical observations because of the growth of electronic messaging to send laboratory orders and test results which are often identified using a health system's internal and typically unique code values. As a result, a receiving care system can-

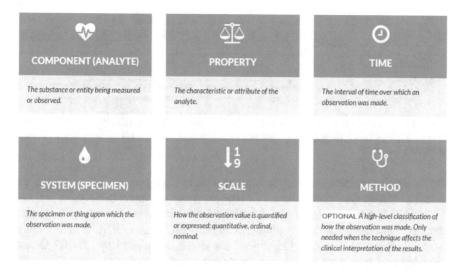

Fig. 7.15 A LOINC name can contain these six components or parts, separated by colons. (Courtesy LOINC)

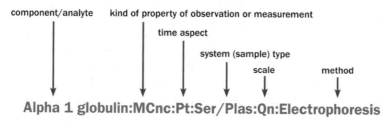

Fig. 7.16 A fully specified LOINC name for a laboratory test indicates the component/analyte, the property analyzed or measured, the time over which this observation or measurement took place, the type of sample, the scale of the result and the method used to obtain it. (Courtesy LOINC)

not fully "understand" and properly file the results they receive without considerable effort. LOINC provides a standard that everyone can use and understand. LOINC is open source and much of the work is done by volunteers. You will find references to LOINC in a number of FHIR Resources. Each code is a number with up to seven digits. While the codes themselves are deceptively simple, their name contains important details in its five or six components, as shown here in Fig. 7.15.

These components, or parts, are separated by colons. In the example in Fig. 7.16, the first part – the component or analyte, the substance of interest – is Alpha 1 Globulin, a protein that is a marker for inflammation. We can see in Fig. 7.15 that the sixth part, the method, is optional. In the example shown in Fig. 7.16 it specifies the method used to determine the level as electrophoresis. The fourth or specimen part indicates that the test was performed on a blood serum or plasma sample. Test results with a unit of measure of mass in the numerator and a volume (like mg/dl) in

Fig. 7.17 An example of the fully specified LOINC description for LOINC code 59459-8 the Mental Status clinical finding shows the use of an outside scale to assess the patient's status. (Courtesy LOINC)

the denominator have mass concentration (MCnc), as you see in the fifth part. Finally, Pt in the third part indicates that the measurement represents a point in time.

LOINC is a bit unusual in that it codes for two domains. The example we just discussed is a laboratory text result which is the domain that LOINC is most widely used for. However, clinical observations are also encompassed by LOINC. Given the wide variety of possible clinical observations and the fact that these are natural phenomenon that aren't as well understood as manmade laboratory tests, this is arguably a more complex challenge than laboratory tests.

Figure 7.17 illustrates an example of LOINC code 59459-8 for Mental Status. It also illustrates the inclusion of an outside assessment instrument as what would be called a 'value set' in FHIR. This is a widely used technique in LOINC.[13] In this example it's the Morse Fall Scale for identifying fall risk factors.[14] Here the Property part of the name is 'Find', an abbreviation for a clinical finding or observation. The scale is 'Ord', an abbreviation for ordinal meaning the value is from a structured list of possibilities. The method is the Morse Fall Scale which is itself external to LOINC but the results of this hypothetical assessment are shown in the figure.

7.9   National Drug Codes (NDC)

The National Drug Code (NDC) is a US-specific standard for medications maintained by the US Food and Drug Administration (FDA). As shown in Fig. 7.18, it consists of a simple 10 digit, 3-segment number. The first segment indicates that the manufacturer/labeler/vendor is Pfizer Consumer Products. The second indicates that the product is Advil. The third part indicates that the packaging is 24 tablets in a bottle. The same medication can have many NDC codes particularly if its patent has expired so it can be produced by many manufacturers.

---

[13] https://danielvreeman.com/proper-use-of-loinc-question-codes-with-assessment-instrument-methods/

[14] http://www.networkofcare.org/library/Morse%20Fall%20Scale.pdf

Fig. 7.18 An example of
an NDC code for Pfizer
Consumer Healthcare's
Advil in a 24-tablet bottle.
(FDA)

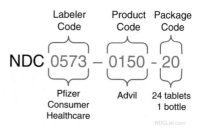

7.10 RxNorm

Pharmacy was an early target for health information technology. The author worked
on one of the first ambulatory pharmacy systems during the early to mid-1970s.[15] As
a result, computer-based codes for medications were also an early development. As
the use of commercial pharmacy software grew, companies created proprietary coding systems for the many aspects of medications, including clinical issues such as
drug-drug interactions, not covered by the simple NDC product codes. Starting in
2001, the National Library of Medicine (NLM) created RxNorm to reconcile these
commercial codes. Each medication has a unique RxNorm RXCUI number of up to
eight digits that can be used to retrieve a great deal of information as you will see in
the exercise that follows. Some FHIR Resources reference RxNorm codes.

7.11 SNOMED Clinical Terms (SNOMED-CT)

The development of SNOMED Clinical Terms traces its roots to a project begun in
the 1960s at National Institutes of Health (NIH) to use natural language processing
(NLP) to machine code pathologists' free text dictated notes. Conceived and headed
by Dr. Arnold W Pratt the first director of the National Institute for Health's Division
of Computer Resources and Technology (now called the NIH Center for Information
Technology) the project developed the Systematized Nomenclature of Pathology
(SNOP). From the outset, SNOP was an ontology representing relationships among
its concepts. By the early 1970s the system was working successfully on an IBM
7094 mainframe computer.[16]

Starting in the mid 1970s under the leadership of Dr. Roger Cote at the University
of Sherbrooke and Dr. David J. Rothwell at the Medical College of Wisconsin, the
College of American Pathologists expanded SNOP to create the Systematized
Nomenclature of Medicine (SNOMED) to meet the growing needs of medicine.

---

[15] http://www.japha.org/article/S0003-0465(16)33478-4/abstract

[16] H Graepel, P & E Henson, D & W Pratt, A. (1975). Comments on the Use of the Systematized
Nomenclature of Pathology. Methods of information in medicine. 14. 72–5. https://doi.
org/10.1055/s-0038-1636818.

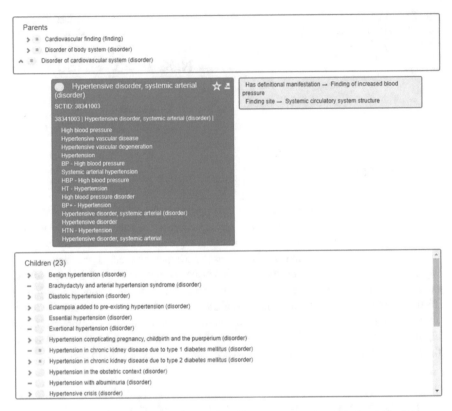

Fig. 7.19 The official SNOMED-CT browser displays the disorder hypertension with all its synonyms, its position in the hierarchical ontology and its 'children' which are more specific forms of the disease. (Courtesy SNOMED International)

Today, it is called SNOMED Clinical Terms and is maintained by SNOMED International. SNOMED-CT is huge and complex and currently has over 311,000 concepts with over 1.3 million relationships among them. Given its scope, you may find references to SNOMED-CT in many types of FHIR Resources.

Concepts are the basic component of SNOMED-CT and have a unique nine-digit SNOMED CT Identifier (SCTID). Each SNOMED CT concept also has a unique, human-readable Fully Specified Name (FSN). Figure 7.19 is from the US edition of the official SNOMED-CT browser[17] and shows the results of a search for the term 'hypertension'. It gets over 700 matches and the first of these – 'hypertension' – with a FSN of 'Hypertensive disorder, systemic arterial (disorder)' is selected. We will just call it hypertension in what follows. Its SCTID is 38341003. Below it you see that SNOMED-CT recognizes 14 synonyms for this disorder. This list could be quite useful in processing text notes or problems on a structured list that might contain one or more of these alternative terms. You can also immediately see the hier-

---

[17] http://browser.ihtsdotools.org/

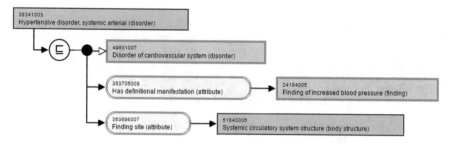

Fig. 7.20 A diagrammatic view of the SNOMED-CT disorder hypertension reveals its clinical manifestations and its location in the body. (Courtesy SNOMED International)

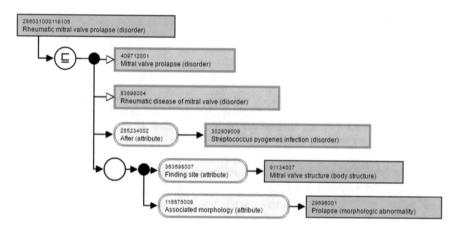

Fig. 7.21 A diagrammatic view of the SNOMED-CT disorder rheumatic mitral valve prolapse reveals its cause (rheumatic disease of the mitral valve as a result of a streptococcus pyogenese infection), its location (mitral valve) and the exact associated physical finding (prolapse of the valve). (Courtesy SNOMED International)

archical nature of SNOMED-CT. Hypertension is a child of 'Disorder of cardiovascular system (disorder)' that is a child of 'Disorder of body system' and so on all the way to the top of the hierarchy. Each of the left arrows in the hierarchy can be expanded so we're only seeing a part of the hierarchy here. Any intermediate level of this hierarchy could be useful in grouping patients for analysis. Hypertension, in turn, has 23 more specific sub-disorders, called children. These could also be useful for grouping patients for analysis.

The browser also provides a diagrammatic hierarchical view shown in Fig. 7.20. A careful look at this shows how SNOMED-CT explicitly reveals important computable clinical relationships. Absent this, how might a computer know that hypertension is associated with an increased blood pressure?

Figure 7.21 is a second, more clinically interesting example, for prolapse of the mitral value due to a rheumatic disease. A physician would know that the root cause was a streptococcal infection that triggered the body to attack this heart valve

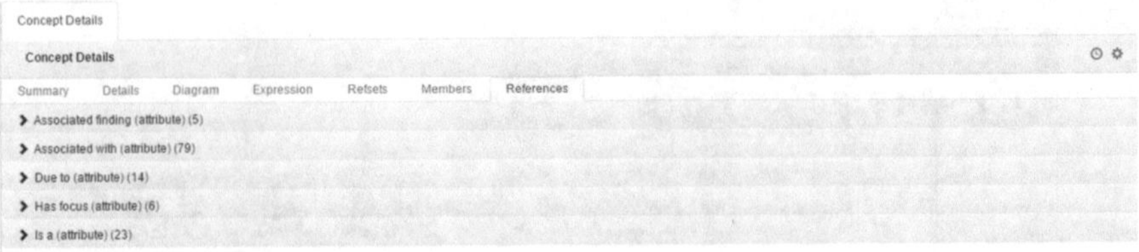

Fig. 7.22 The references view provides many additional clinical details grouped into five headings. (Courtesy SNOMED International)

Fig. 7.23 Expanding the 'Associated With' references group shows that hypertension is linked to kidney, heart and neurological disorders. You may recall that in our discussion of chronic disease we mentioned that more serious disorders can occur if the diseases aren't well controlled and this make those associations explicit. (Courtesy SNOMED International)

because of the similarity of some of its surface proteins to those of the bacteria but how would a computer know this? SNOMED-CT makes it explicit.

Finally, Fig. 7.22 presents a references view that provides a great deal of potentially useful information in the form of relationships between the current concept and others. You can see that they are in five groups. We will briefly discuss only the first two. 'Associated Findings' are clinical observations that might occur with hypertension such as a family history of the disease. 'Associated With' are clinical findings that hypertension patients may have.

Figure 7.23 expands the 'Associated With' group to show, for example, that hypertension patients might have one or more of a number of kidney diseases that can cause hypertension. Again, this is potentially very useful information for analysis of clinical data.

SNOMED-CT is quite a bit more complex than presented here in this brief overview. Nevertheless, it should be clear that it can be a very valuable tool for usefully grouping and analyzing patients and their medical conditions.

7.12   Recap

You now have an overview of how standards have evolved over the years in terms of their structure, purpose and technology. You are also now familiar with most of the key data standards. You should think about the difference between simpler classifications such as NDC, CPT and ICD (prior to version 10) and a complex ontology, such as SNOMED-CT. The standards community has struggled over the tension between "perfection" – a standard that can represent medicine in all its detail – and "practicality" – standards that can actually be deployed and used in the real world. This dichotomy extends equally to the topic of our next chapter, interoperability standards, and it eventually led to FHIR, the subject of the chapter after that.

Chapter 8
Pre-FHIR Interoperability and Clinical Decision Support Standards

8.1 Introduction

In the last chapter, we looked at the evolution of standards and examined the key health data standards. For use in actual patient care, standardized clinical data, typically along with other non-standardized data such as free text notes, must be packaged into a useful and usable form and sent using widely accepted data sharing standards and approaches. Many people feel that these standards and approaches will increasingly be FHIR APIs but FHIR is still being developed and the first 'normative' version isn't expected until late 2018 so for some time into the future earlier standards will continue to be used. We will look at FHIR in the next chapter, while in this one we will look at the standards for packaging and sharing data that predate it.

We will also look at Arden, a standard for the representation of clinical logic for decision support that was arguably ahead of its time. Arden is interesting and discussing it in some detail should serve to set the stage for later discussions of the key role that FHIR can play in making decision support tools for improved patient care easier to implement and use.

8.2 HL7 Evolution

What follows is an overview of many of the key components of HL7 with a particular emphasis on those that support data sharing for care coordination and patient engagement. It is important to keep in mind that, over the years, HL7 has expanded greatly from its original messaging mission and from the basic, but pragmatic, approach taken in V2 message standards development. The results of these newer efforts have not been widely adopted and some would now argue that this is because HL7's later standards were overly complex and too difficult and expensive to implement in practice.

© Springer International Publishing AG, part of Springer Nature 2018
M. L. Braunstein, *Health Informatics on FHIR: How HL7's New API is Transforming Healthcare*, https://doi.org/10.1007/978-3-319-93414-3_8

8.3 HL7 V2 Versus V3

The last chapter provided a simple comparison of HL7 V2 and V3 messages. We will now look at that in more detail to help us understand the evolution of HL7 standards development.

HL7 V2 is not "Plug and Play". The concept was just getting started back then in the hardware world and certainly was not very common. V2 provides 80% of the interface and a framework to negotiate the remaining 20% on a case-by-case basis. Thus, developing an HL7 interface between any two systems requires coding and has a certain degree of customization. Part of the reason why, is that V2 was not based on a standard model.

A key goal of HL7 V3 development was to get as close to "Plug and Play" as possible. As we have discussed, it uses the Reference Information Model (RIM) in an effort to provide consistency across the standard.

8.4 Reference Information Model (RIM)

The goal of the HL7 Reference Information Model or RIM is to document the actions taken to treat a patient. A request or order for a test is an action. The reporting of the test result is an action. Creating a diagnosis based on test results is an action. Prescribing treatments based on that diagnosis is an action. RIM defines the semantics of a common set of administrative, financial and clinical concepts in order to describe these actions and foster interoperability.

Figure 8.1 is a simple diagrammatic representation of the four core RIM concepts or backbone classes (entity, role, participation and act) and their relationships. Every happening is an Act that is either being done, has been done, can be done or has been requested or ordered. Examples include clinical observations, medication administration, medical procedures, or patient encounters.

Act relationships represent connections between acts such as composition, preconditions, revisions and support. Participation defines the context for an Act such as author, performer, subject, or location. The participants have Roles such as patient, provider, practitioner, specimen or healthcare facility. Entities such as persons, organizations, material, places or devices play these roles.

How can a single act class represent all of the elements of a clinical action – their definition, requesting or ordering them or reporting the result? This is the role of the Act "mood" code that specifies whether the Act is an activity that is defined, is an event, has been requested or ordered, or is promised or is the subject of a future appointment.

Figure 8.2 is an example of a patient registration message in both the V2 and V3 formats. You will not find anything related to RIM in the V2 message at the top. Lower down, in the V3 version **<processingModeCode" code = "T"/>** is highlighted. We have said that FHIR borrows value sets from RIM and to illustrate that on both the HL7 RIM and FHIR web sites you can find the table shown in Fig. 8.3

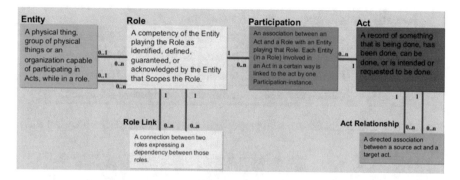

Fig. 8.1 Health Level Seven (HL7) RIM uses an object-oriented modeling approach derived from four main classes: Entity, Role, Participation and Act. (Courtesy HL7)

```
MSH|^~\&|EPIC|EPICADT|SMS|SMSADT|199912271408|CHARRIS|ADT^A04|1817457|D|2.5|
PID||0493575^^^2^ID 1|454721||DOE^JOHN^^^^|DOE^JOHN^^^^|19480203|M||B|254 SOME ST.
^^NEWTOWN^OH^44124^USA||(332)123-4567|||M|NON|400003403~1129086|
NK1||ROE^MARIE^^^^|SPO||(216)123-4567||EC|||||||||||||||||||||||||||
PV1||O|168 ~219~C~PMA^^^^^^^^^||||277^ALLEN CARSTON JR.^CARLY^^^^||||||||| ||
2688684|||||||||||||||||||||||||199912271408||||||002376853
```

```xml
<?xml version="1.0" encoding="UTF-8"?>
<PRPA_IN101001UV01 ITSVersion="XML_1.0" xmlns="urn:hl7-org:v3"
    xmlns:xsi="http://www.w3.org/2001/XMLSchema-instance">
  <id extension="3948375" root="2.16.840.1.113883.19.10.700363.2288"/>
  <creationTime value="20060501140010"/>
  <versionCode code="NE2006"/>
  <!-- Interaction is a notification of a person registration -->
  <interactionId extension="PRPA_IN101001UV01" root="2.16.840.1.113883.1.6"/>
  <processingCode code="P"/>
  <processingModeCode code="T"/>
  <acceptAckCode code="ER"/>
  <receiver>
    <device>
     <id extension="922" root="2.16.840.1.113883.19.9"/>
     <name>Master MPI</name>
      <asAgent>
        <representedOrganization>
          <id extension="1002003" root="2.16.840.1.113883.19.200"/>
          <name>Alpha Hospital</name>
        </representedOrganization>
      </asAgent>
    </device>
  </receiver>
  <sender>
    <device>
      <id extension="1" root="2.16.840.1.113883.19.9"/>
      </device>
```

Fig. 8.2 A message of patient registration in both HL7 V2 messaging format (top) and V3 format (bottom) clearly illustrates the use of the RIM in V3. (Courtesy HL7)

that explains the possible values of this code – its value set. 'T' indicates 'Current Processing' so that is the status of this patient's registration.[1,2]

The state of processing of an administrative task is but one example of the great level of detail encompassed by RIM. To get a further feel for this Fig. 8.4 highlights

---

[1] http://www.vico.org/HL7_V3_CD1_2012/Edition2012/infrastructure/vocabulary/ProcessingMode.html

[2] https://www.hl7.org/fhir/v3/ProcessingMode/vs.html

Level	Code	Display	Definition
1	A	Archive	Identifies archive mode of processing.
1	I	Initial load	Identifies initial load mode of processing.
1	R	Restore from archive	Identifies restore mode of processing.
1	T	Current processing	Identifies on-line mode of processing.

Fig. 8.3 This value set for the HL7 RIM processingMood is also used in FHIR. (Courtesy HL7)

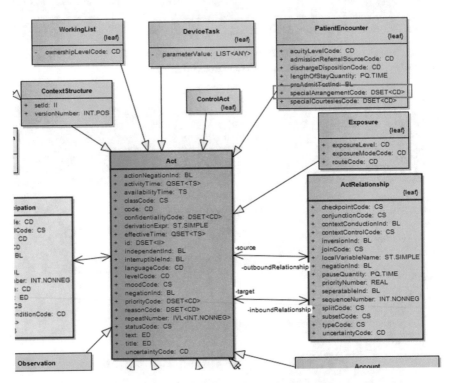

Fig. 8.4 The specialArrangementCode leaf of the PatientEcounter provides for special things a patient might need on arrival for that encounter illustrating the enormous level of detail encompassed by RIM. (Courtesy HL7)

the **specialArrangementCode** for a Patient Encounter. Note that **PatientEncounter** is a leaf of the Act box meaning it is a particular example or instance.

The value of the **specialArrangementCode** is DSET indicating there is a discrete set of possible values (a value set) for this field. Once again, as you see in Fig. 8.5, the FHIR site lists this value set that codes for special things a patient might need at arrival for an encounter. The possible values are "wheel" for wheelchair, "add-bed" for additional bedding, "int" for interpreter, "att" for attendant and "dog" for guide dog.

Code	Display	Definition
wheel	Wheelchair	The patient requires a wheelchair to be made available for the encounter.
add-bed	Additional bedding	An additional bed made available for a person accompanying the patient, for example a parent accompanying a child.
int	Interpreter	The patient is not fluent in the local language and requires an interpreter to be available. Refer to the Patient.Language property for the type of interpreter required.
att	Attendant	A person who accompanies a patient to provide assistive services necessary for the patient's care during the encounter.
dog	Guide dog	The patient has a guide-dog and the location used for the encounter should be able to support the presence of the service animal.

Fig. 8.5   The RIM value set for the specialArrangementCode of an encounter. (Courtesy HL7)

8.5   RIM and FHIR

You can find the ValueSets for many RIM codes on the FHIR site because FHIR builds on existing standards including the RIM. However, there are exceptions.

Scope: It should be clear by now that the goal of RIM is comprehensive representation of the health care domain. FHIR resources represent only those data elements that are expected to be used by "most" implementations. FHIR uses an 80% rule – if at least 80% of standards contributors say they will need it, a resource will support a particular data element, as part of the core FHIR specification. You should read the post by Grahame Grieve on this often misunderstood FHIR design principle.[3]

A specific sub-domain can add other elements, as needed, using FHIR extensions. FHIR Profiles both constrain resources (e.g. such as by specifying what values from a value set are allowed) for a particular use case and define extensions appropriate to special subdomains.

Source of Data Elements: All data elements in HL7 V3 instances come from either the RIM or data types developed by the International Organization for Standardization (ISO). In FHIR, this is true of most, but not all, resources and data type elements. Some FHIR resources deal with content that is outside the RIM's scope and, in a few instances, FHIR adjusts data types to accommodate issues not yet supported in RIM.

---

[3] http://www.healthintersections.com.au/?p=1924

Nuance: As you have seen, RIM attempts to convey the meaning of instances through attributes like the Mood Code. FHIR codes are generally limited to attributes with a rather concrete business meaning.

Finally, in FHIR a ValueSet is a resource that can be sent as part of an instance just like any other piece of data. As a result, although FHIR uses RIM data elements, it is possible to implement FHIR with absolutely no knowledge of the HL7 RIM. This greatly facilitates rapid development.

Before moving on to some of the things RIM is used for, it is important to emphasize again that the goal is interoperability and, in support of that, RIM can be used both for HL7 V3 messages and for clinical documents constructed, at least in part, from the data in them.

8.6   Clinical Document Architecture Uses RIM

The Clinical Document Architecture or CDA defines HL7 V3 RIM-based documents assembled from administrative and clinical data for particular purposes. You saw one of these if you did the HealthVault CCD activity. You should also recall that a consolidation effort produced the newer C-CDA standard. Figure 8.6 is the first part of the CCD you used in the HealthVault activity. The instances where it refers to the RIM are highlighted.

I leave it to you to consider the balance between value, complexity and ease of implementation that this reliance on RIM entails.

8.7   C-CDA Templates

C-CDA documents are assembled from templates, essentially reusable XML components. Templates are defined at the document, section or data entry level. These correspond conceptually to the parts of a paper form where the document as a whole consists of sections, each of which consists of fields into which data is recorded.

Figure 8.7 is the same CCD we just looked at in Fig. 8.6 but the two references to templates are also highlighted but in a darker color.

There are C-CDA template guides posted including a graphically well-designed example from HL7.[4] Templates have an OID, a globally unique ISO identifier. In Figs. 8.6 and 8.7, the first one has an OID of 2.16.840.1.113883.10.20.22.1.1. The second differs from the first only in the last digit. Both of these are document level templates. The first defines a document header for use in the US domain. This serves a purpose quite similar to the FHIR Profiles in that it constrains the document to US specifications. In Fig. 8.8 (from the template guide referenced earlier) you see that it specifies that there must be exactly one realm code and its value must be "US".

---

[4] http://ccda.art-decor.org/ccda-html-20150727T182455/index.html

```
<?xml version="1.0" encoding="utf-8"?>
<?xml-stylesheet type="text/xsl" href="CDA.xsl"?>
<!--Copyright EXACTDATA, LLC. ALL RIGHTS RESERVED.-->
<!--This file contains synthetic data depicting a fictional
health summary. Any resemblance to a real person is
coincidental.-->
<ClinicalDocument xmlns:xsi="http://www.w3.org/2001/XMLSchema-
instance" classCode="DOCCLIN" moodCode="EVN" xmlns="urn:hl7-
org:v3">
     <realmCode code="US" />
     <typeId root="2.16.840.1.113883.1.3"
extension="POCD_HD000040" />
     <templateId root="2.16.840.1.113883.10.20.22.1.1" />
     <templateId root="2.16.840.1.113883.10.20.22.1.2" />
     <id root="EFCE504E-60F1-48F0-8EF8-6A546AEC8C4C" />
     <code code="34133-9" codeSystem="2.16.840.1.113883.6.1"
codeSystemName="LOINC" displayName="Summarization of Episode
Note" />
     <title mediaType="text/plain" representation="TXT">
Continuity of Care Document</title>
     <effectiveTime value="201408261600-0400" />
     <confidentialityCode code="N"
codeSystem="2.16.840.1.113883.5.25" displayName="normal" />
     <languageCode code="en-US" />
     <recordTarget typeCode="RCT" contextControlCode="OP">
          <patientRole classCode="PAT">
               <id root="2.16.840.1.113883.4.1"
extension="527-67-0000" />
               <id root="2.16.840.1.113883.3.441.1.50.300011.51"
extension="527670000-01" />
```

Fig. 8.6 A CDA document contains many references to the RIM. (Courtesy HL7)

It is more interesting to look at the second template (Fig. 8.9) that specifies constraints for a Continuity of Care Document (CCD) in conformance with Stage 1 Meaningful Use. Here, you see that it specifies the sections that must be contained in the CCD so that it adequately supports use cases such as Transitions of Care. Each of these refers to a Section template and the HL7 site referenced earlier provides links to these more granular templates.

If you first click on the line to the Medications Section template and, from it, click on the **MedicationActivity** link you get to the Data Entry level template shown in part in Fig. 8.10. Note that it, in turn, links to specific ValueSets such as the Medication Route, Body Site, Units of Measure or Rate Quantity. Clicking on any of these links brings up a page listing the available values which, in the case of Route, came from an external source, the US Food and Drug Administration (FDA). It also links to specialized Data Entry templates. Not shown here is the Medication Information Entry Level template that could even contain a pre-specified product

```
<?xml version="1.0" encoding="utf-8"?>
<?xml-stylesheet type="text/xsl" href="CDA.xsl"?>
<!--Copyright EXACTDATA, LLC. ALL RIGHTS RESERVED.-->
<!--This file contains synthetic data depicting a fictional
health summary. Any resemblance to a real person is
coincidental.-->
<ClinicalDocument xmlns:xsi="http://www.w3.org/2001/XMLSchema-
instance" classCode="DOCCLIN" moodCode="EVN" xmlns="urn:hl7-
org:v3">
    <realmCode code="US" />
    <typeId root="2.16.840.1.113883.1.3"
extension="POCD_HD000040" />
    <templateId root="2.16.840.1.113883.10.20.22.1.1" />
    <templateId root="2.16.840.1.113883.10.20.22.1.2" />
    <id root="EFCE504E-60F1-48F0-8EF8-6A546AEC8C4C" />
    <code code="34133-9" codeSystem="2.16.840.1.113883.6.1"
codeSystemName="LOINC" displayName="Summarization of Episode
Note" />
    <title mediaType="text/plain" representation="TXT">
Continuity of Care Document</title>
    <effectiveTime value="201408261600-0400" />
    <confidentialityCode code="N"
codeSystem="2.16.840.1.113883.5.25" displayName="normal" />
    <languageCode code="en-US" />
    <recordTarget typeCode="RCT" contextControlCode="OP">
        <patientRole classCode="PAT">
            <id root="2.16.840.1.113883.4.1"
extension="527-67-0000" />
            <id root="2.16.840.1.113883.3.441.1.50.300011.51"
extension="527670000-01" />
```

Fig. 8.7 To assure consistency to the standard C-CDA documents are built from reusable templates. These can be at the document level, section level or date entry level. Each has a unique OID (beginning with 2.16…). (Courtesy HL7)

Item	DT	Card	Conf	Description
cda:ClinicalDocument				
└ cda:realmCode		1..1	M	SHALL contain exactly one [1..1] realmCode="US" (CONF:16791).
└ cda:typeId		1..1	M	SHALL contain exactly one [1..1] typeId (CONF:5361).

Fig. 8.8 This simple document level template simply specifies that it is for use only in the US realm using the CDA realmCode "<realmCode code="US"/>". (Courtesy HL7)

strength, product form, or product concentration ("amoxicillin 400 mg/5 mL suspension"). Clicking on its Medication Clinical Generic Drug link shows that the specification must be made using RxNorm.

Fig. 8.9 The second document level template specifies the sections that must be contained in a US Continuity of Care Document (CCD) if it is to meet the requirements of Meaningful Use. Each link is to the corresponding section level template. (Courtesy HL7)

Fig. 8.10 Many data fields are limited to defined value sets that might come from external sources such as the FDA or RxNorm. (Courtesy HL7)

8.8   Clinical Decision Support

Clinical decision support (CDS) is an early, important and particularly interesting, domain within health informatics. Its purpose is to provide clinicians, patients and others with knowledge and personalized information, intelligently filtered or presented at appropriate times, to enhance health and health care. It is a critical component of the IOM's vision of a Learning Health System since it is the vehicle for feeding knowledge obtained from the care of prior patients back to providers caring for current patients.

According to ONC, "CDS tools include computerized alerts and reminders to care providers and patients; clinical guidelines; condition-specific order sets; focused patient data reports and summaries; documentation templates; diagnostic support, and contextually relevant reference information, among other tools."

We will discuss three key CDS projects to illustrate how early researchers recognized CDS as an opportunity and the challenges to successfully implementing it in practice. This discussion will set the stage for appreciating how two key capabilities of FHIR and SMART on FHIR are making CDS a far more approachable and clinically useful tool.

The first key capability is facile retrieval of EHR data in a standard format facilitating the integration of FHIR apps into any EHR. This also obviates the need for any redundant data entry by the provider. The second is seamless integration into the providers' workflow making the use of CDS far more efficient and, therefore, more likely.

This discussion will begin with some of the most fascinating stories in the history of health informatics and the vision of some of its pioneers. Space does not allow for a full treatment of this particularly rich domain within the field and this discussion should not be construed as comprehensive since it does not describe all the projects worthy of inclusion.

8.9   Dr. Homer Warner's HELP System

Beginning in the mid-1950s, Dr. Homer Warner, one of the founders of health informatics, began using computers for decision support in cardiology at LDS Hospital (now Intermountain Healthcare) in Salt Lake City. In the 1970s, Dr. Warner and his colleagues created the Health Evolution through Logic Processing or HELP system, one of the first electronic medical record systems, and perhaps the first designed to assist clinicians in decision-making. Intermountain replaced HELP2, the successor to HELP, with the Cerner system in 2015. Dr. Warner died in 2012 and Intermountain named its research center after him. The American Medical Informatics Association (AMIA) awards a cash prize in Dr. Warner's honor each year at its annual symposium.

You can see in Fig. 8.11 that the HELP system's knowledge-based decision processor received data from virtually all areas of the hospital to provide a variety of CDS tools. Note that HELP created what we would call today a longitudinal clinical data repository. It might surprise many in the field today that such a system was functional that long ago.

Expert panels organized much of the clinical knowledge in an early example of knowledge engineering. In fact, in 1997, Dr. Warner published the book *Knowledge Engineering in Health Informatics*.[5] Figure 8.12 is a diagram from that book. In it you see how clinical findings link to a diagnosis. Note the logical rule which states that if findings a or b and any two of the other four findings are present in the same patient, the diagnosis is confirmed.

---

[5] http://www.springer.com/us/book/9780387949017

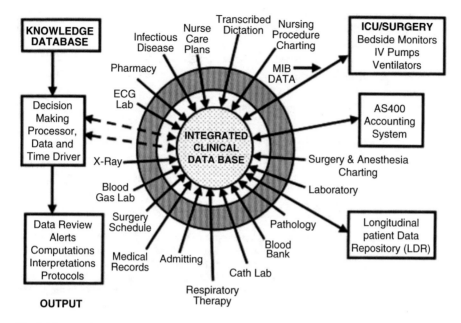

Fig. 8.11 The HELP system brought together data from virtually all areas of the hospital to create an integrated clinical database to drive its decision making processor. (Courtesy Springer)

<u>Lung consolidation</u>
  a. PE shows abnormal chest percussion with dullness
  b. PE shows abnormal pulmonary auscultation with bronchial breath sounds
  c. PE shows abnormal pulmonary auscultation with egophony
     (E-to-A changes)
  d. PE shows abnormal chest palpation with increased vocal
     fremitus
  e. PE shows abnormal pulmonary auscultation with rales
  f. PE shows abnormal pulmonary auscultation with
     pectoriloquy
True if (a or b) and 2 of (c,d,e,f)

Fig. 8.12 A group of clinical findings on physical examination (PE) when combined with logic rules programs the decision making processor to recognize a specific problem, in this case lung consolidation. (Courtesy Springer)

In Fig. 8.13 you see that HELP used Bayesian statistics to relate findings to diagnoses. The first column in this Bayesian Frame is the True Positive (TP), the probability that the finding will be present if the patient has the condition. Thus, it is virtually certain that a patient with iron deficiency anemia will be both anemic and iron deficient. The second column is the False Positive (FP), the probability that a patient with that finding does not have the condition. Thus only 10% of anemic

Iron deficiency anemia (a priori = .075)		
a. Anemia	.999	.10
b. Hypochromic and microcytic RBSs	.85	.07
c. Iron deficiency	.9999	.01
xor		
At risk for iron deficiency	.95	.25
xor		
Chronic blood loss	.95	.10
d. Absolute reticulocyte count		
<50,000	.90	.08
50,000-200,000	.10	.84
≥200,000	.001	.08

Fig. 8.13 This figure shows how HELP uses Bayesian statistics. The first column in this Bayesian Frame is the True Positive (TP), the probability that the finding will be present if the patient has the condition. The second column is the False Positive (FP), the probability that a patient with that finding does not have the condition. The Bayes equation is applied sequentially to these findings where their value is known to calculate the probability that the diagnosis is present in the patient. (Courtesy Springer)

patients have a basis for that anemia other than iron deficiency. The Bayes equation is applied sequentially to these findings where their value is known to calculate the probability that the diagnosis is present in the patient.

8.10 MYCIN

MYCIN is another early example of the use of expert system and artificial intelligence techniques to provide clinical decision support. Specifically, it was an early backward chaining expert system. This technique starts with the goal (is there an antibiotic whose use is supported by the clinical evidence) and works back to the available data to see if it provides the necessary support for a clinical recommendation. MYCIN also provided dosage adjusted for patient's body weight. The software was written in Lisp (one of the oldest programming languages dating back to the late 1950s, the name derives from "LISt Processor") as the doctoral dissertation of a future health informatics pioneer, Edward "Ted" Shortliffe.[6] The name derives from a commonly used family of antibiotic drugs. Interestingly, Shortliffe worked in the same lab that developed DENDRAL, an even earlier expert system to aid chemists in determining the structure of organic molecules.

According to Dr. B.J. Copeland, Director of the Turing Archive for the History of Computing, University of Canterbury, MYCIN worked well: "using about 500 production rules, MYCIN operated at roughly the same level of competence as human specialists in blood infections and rather better than general practitioners."[7]

---

[6] http://www.shortliffe.net/

[7] https://www.britannica.com/technology/MYCIN

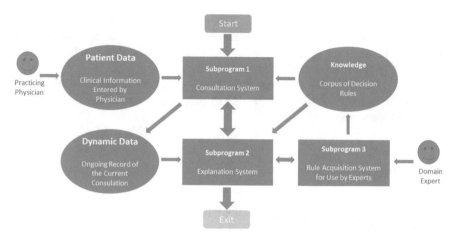

Fig. 8.14 The MYCIN architecture shows that, once again, clinical rules derived from experts inform the expert system. This tedious approach is now rapidly giving way to 'deep learning' which can derive the rules from vast quantities of digital medical record (and other) data. (Courtesy of Dr. Ted Shortliffe)

Despite this impressive performance, MYCIN was never used in routine clinical practice. Beyond concerns expressed about using an expert system in patient care, MYCIN, like virtually all clinical decision support systems, suffered from being stand-alone systems requiring the use of a unique user interface and re-entry of already recorded clinical data. As a result, according to Wikipedia, "a session with MYCIN could easily consume 30 minutes or more—an unrealistic time commitment for a busy clinician."[8]

Figure 8.14 presents a high level view of MYCIN's architectures. It is a simplified representation so while the patient data come from the clinician and the rules come from experts both the clinicians and the experts use the same system. The 'start' box applies to both of them. The rule acquisition system (subprogram 3) is available only to experts, but to identify necessary adjustments to the knowledge base, they use subprograms 1 and 2 to run cases. When they note problems, they then use subprogram 3 to "fix" existing rules or add new ones. The clinician uses only subprograms 1 and 2 in order to get advice. He or she enters patient data while using subprogram 1.

Like HELP, MYCIN depending on some 600 rules derived from expert clinicians. It is worth noting that today this step would likely be replaced with the deep learning tools increasingly able to figure out the rules based on the analysis of vast stores of digital data that are now available in an era of electronic health record systems. However, in a communication with the author, Dr. Shortliffe points out "that there is still a place for knowledge-based guidance of the ML methods. Blind pattern recognition, even with "deep" approaches, can go awry in ways that any expert would see reflect the lack of domain knowledge within the algorithms."

---

[8] https://en.wikipedia.org/wiki/Mycin

In response to a physician query MYCIN provided a ranked list of possible bacteria, the probability of each diagnosis with its confidence in each diagnosis and the reasoning behind each diagnosis (list of questions and rules which led to its ranking). Finally, it recommended a drug treatment.

In 1988 Dr. Shortliffe recognized that the computers on which MYCIN was designed to run were rapidly disappearing so he recorded a set of three videos demonstrating the system.[9,10,11]

8.11   INTERNIST

INTERNIST was an arguably even more ambitious early CDS. It was developed starting in 1974, at the University of Pittsburgh by Dr. Jack Myers, an internal medicine physician, and Dr. Harry E Pople, a computer scientist and pioneer in artificial intelligence. Health informatics is not often the first place a new computer science technique appears but, according to volume 64 of the Encyclopedia of Library and Information Science, "Applying Abduction to artificial intelligence problems began with Harry Pople and his system, INTERNIST."[12]

The 1985 book *Logic of Discovery and Diagnosis in Medicine* describes the program as "an AI partial simulation of Dr. Myer's clinical reasoning using his own internal knowledge base".[13] Once again clinical rules derived in this case from one practitioner were the basis for the expert system.

The goal of INTERNIST was to make the appropriate diagnosis in a given clinical situation. In the 1985 book Dr. Pople describes the methods used. They are summarized here.

He explains why either a Bayesian approach or a branching logic approach (essentially a network) that starts with a presenting complaint, as shown in Fig. 8.15, are inadequate to deal with the real world of clinical diagnosis where new or conflicting data may arrive too late to assist with an interim decision, and it may even arrive in random order.

INTERNIST contained a knowledge base consisting of some 500 disease entities, organized into categories by organ system, and their over 3,000 clinical manifestations. The diagram of liver diseases and their related conditions and manifestations in Fig. 8.16 illustrates these relationships and is from a chapter by Dr. Pople in the 1982 book *Artificial Intelligence in Medicine*.[14] Each disease entity had a corresponding list of manifestations weighted on a scale of 1–5 (the weight-

---

[9] https://www.youtube.com/watch?v=a65uwr_O7mM&t=5s

[10] https://www.youtube.com/watch?v=ppkg4mQIgXw&t=6s

[11] https://www.youtube.com/watch?v=bro6fkDxCUE

[12] https://tinyurl.com/y88anayb

[13] https://tinyurl.com/yb4hcr5r

[14] https://tinyurl.com/yasrmqe8

Fig. 8.15 A branching logic approach starting with the presenting complaint was determined to be inadequate for INTERNIST to deal with real world clinical diagnosis because the needed data might arrive too late or in random order. (*Artificial Intelligence in Medicine*, 1982)

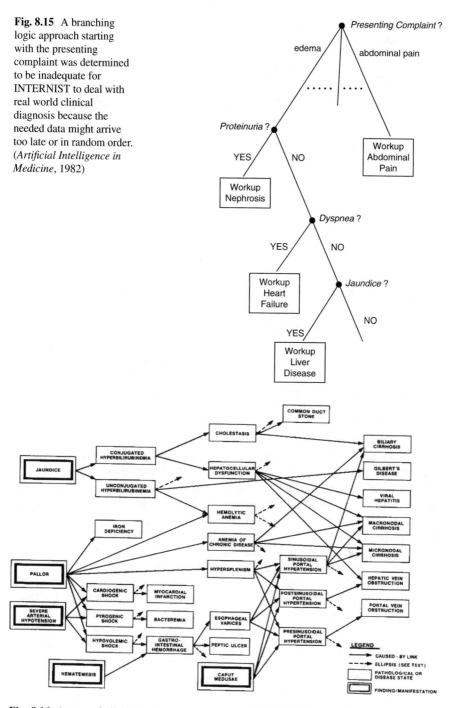

Fig. 8.16 A map of clinical relationships and from INTERNIST links various manifestations, their interrelationships and their association with a group of liver diseases shown on the right. (*Artificial Intelligence in Medicine*, 1982)

```
PURSUING:   DISSEMINATED INTRAVASCULAR COAGULATION
ACUTE

PLEASE ENTER FINDINGS OF COAGULATION TEST (S)
*GO

ANTITHROMBIN III LESS THAN 50 PERCENT OF NORMAL?
N/A

ETHANOL GEL TEST POSITIVE?
N/A

PROTAMINE PARACOAGULATION TEST POSITIVE?
N/A

BLOOD INCOAGULABLE?
NO
```

Fig. 8.17 Once it developed its list of possible diagnoses based on how well they were supported by the clinical findings INTERNIST asked the clinician user questions to further discriminate and attempt to reach a final diagnosis. (*Artificial Intelligence in Medicine*, 1982)

ings aren't shown in the figure) based on the frequency of their occurrence in that disease. Conversely, a scale of 0–5 reflected the strength of the association of manifestations to diseases. Finally, there was representation of causal, temporal and other relationships among entities. This structure is in many ways suggestive of SNOMED-CT, whose predecessor, SNOP, was by then in operation at NIH but there is no way to know now if the INTERNIST team was aware of it.

INTERNIST ranked disease entities that explain any or all of a patient's findings reflecting their goodness of fit to the data. It then formulated questions for the physician, as shown in Fig. 8.17, to discriminate among the equally ranked entities. In this example, the computer is considering the diagnosis of Acute Disseminated Intravascular Coagulopathy (DIC) a dangerous, potentially life-threatening condition in which blood clots form in the small blood vessels. Much like an attending physician might do on rounds with physicians in training, it is asking for the results of the pertinent laboratory tests to confirm or rule out the diagnosis. Once a diagnosis was considered identified it was added to the patient's problem list; its manifestations were considered accounted for; and the process repeated.

While INTERNIST worked, in complex cases it often considered inappropriate possibilities and wasted clinicians' time answering questions with respect to them. To overcome this, INTERNIST-II used a heuristic that recognized that certain manifestations are distinctively characteristic of a disease or category of diseases. Jaundice, which is prominent in Fig. 8.16, is highly suggestive of liver disease so the initial focus in a patient with this symptom should be on this disease category, as it would be if a physician were making the diagnosis.

INTERNIST-II also introduced a multi-problem generator, essentially a search process beginning globally with 'health problem' as shown in Fig. 8.18. It terminated when at least one problem hypothesis accounted for all clinical findings.

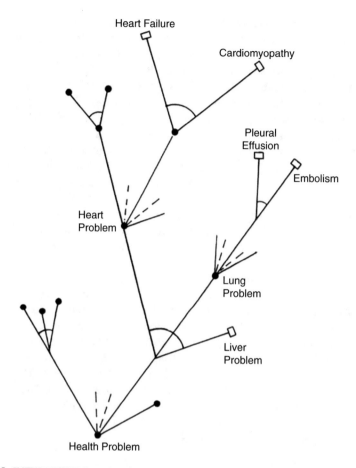

Fig. 8.18 INTERNIST-II introduced a multi-problem generator, essentially a search process beginning globally with 'health problem', as shown here. It terminated when at least one problem hypothesis accounted for all clinical findings. (*Artificial Intelligence in Medicine*, 1982)

However, the hypotheses might or might not specify specific diseases. If there were sufficient cues in the data, further hypothesis reduction could reduce the possibilities to specific problems. Finally, it further reduced the problem list by exploiting its database of causal, temporal and other relationships, yet another process analogous to what actual human physicians do.

8.12   Arden: A Standard for Medical Logic

Systems such as HELP and INTERNIST are impressive, but each employed their own approach to the representation of clinical knowledge and logic. Developing these systems was a substantial effort, far too formidable for most institutions that might lack either the clinical or computer science expertise to take on such a project. Clearly, it would be preferable to implement existing knowledge and clinical logic in institutions other than the one in which they were developed. Doing so inevitably means finding a way to connect their 'clinical logic' to other EHRs.

The HL7 Arden Syntax for Medical Logic Modules (MLMs) is an American National Standards Institute (ANSI) approved language for encoding medical knowledge and representing and sharing that knowledge. The primary goal is sharing of medical logic across EMRs for Clinical Decision Support (CDS).

A first draft of the standard was prepared in 1989 at a meeting at the Arden Homestead in Harriman, NY. The homestead is part of Columbia University and Dr. George Hripcsak, now chair of its Department of Biomedical Informatics, led the effort. The American Society for Testing and Materials (ASTM) adopted the Arden standard and published it in 1992. Later, the standard was integrated into Health Level Seven (HL7) which published Arden Syntax version 2.0 in 1999 and has been hosting its further development ever since. The latest version, 2.10, was published in November 2014.

Arden's design borrowed from the HELP system that eventually converted to the syntax. Arden also borrowed from the Regenstrief Institute's CARE language, a rule-based syntax that could generate reminders or retrieve patient records based on pre-specified criteria.[15] CARE, and the Regenstrief Institute Record System (RMRS), are other early and important innovations in health informatics. RMRS development, begun in 1972 using data from 35 of Dr. Charles Clark's diabetes patients, was headed by another health informatics pioneer, Dr. Clement J. McDonald, who we mentioned earlier as the organizer of the LOINC development effort. RMRS was one of the first EMR systems to offer rule-based CDS. Three hospitals on the Indiana University Medical Center campus and more than 30 Indianapolis clinics currently use RMRS.[16]

8.13   Arden Explained

The basic building blocks of Arden clinical decision support rules are Medical Logic Modules (MLM) each of which contains sufficient knowledge to make a single clinical decision. As illustrated in Fig. 8.19, MLMs use the Backus-Naur Form (BNF), a notation for context-free grammars, otherwise used to describe the syntax of computer programming languages, document formats, instruction sets and communication protocols.

Each MLM can have four categories of information: maintenance, library, knowledge and resources. The examples in Figs. 8.20 and 8.21 are from Dr. Hripcsak. Figure 8.20 illustrates two categories. The category names and the key parts of the logic are highlighted. The purpose of the Library category of this MLM is to check

---

[15] McDonald, CJ et al. Implementing Healthcare Information Systems, Orthner, Helmuth F., Blum, Bruce (Eds), Springer, 1989, p 82.

[16] http://www.sciencedirect.com/science/article/pii/S138650569900009X#BIB25

<expression> - represents the non-terminal expression

"IF" – represents the terminal if, iF, If, or IF

":=" – represents the terminal :=

::= - is defined as

/*...*/ - a comment about the grammar

| - or

Fig. 8.19 Arden MLMs use the Backus-Naur Form (BNF), a notation for context-free grammars, otherwise used to describe the syntax of computer programming languages, document formats, instruction sets and communication protocols. (Courtesy Dr. George Hripcsak)

```
library:
    purpose:
            When a penicillin is prescribed, check for an allergy. (This MLM
            demonstrates checking for contraindications.);;
    explanation:
            This MLM is evoked when a penicillin medication is ordered. An
            alert is generated because the patient has an allergy to penicillin
            recorded.;;
    keywords: penicillin; allergy;;
    citations: ;;
knowledge:
    type: data-driven;;
    data:
            /* an order for a penicillin evokes this MLM */
            penicillin order := event {medication order where
                                       class = penicillin};
            /* find allergies */
            penicillin allergy := read last {allergy where
                                             agent class = penicillin};
            ;;
    evoke:
            penicillin_order;;
    logic:
            if exist(penicillin allergy) then
                conclude true;
            endif;
            ;;
    action:
            write "Caution, the patient has the following allergy to penicillin documented:"
             || penicillin_allergy;;
    urgency: 50;;
end:|
```

Fig. 8.20 The purpose of the Library category of this MLM is to check each new penicillin prescription for a penicillin allergy. In the Knowledge category the triggering event is that the newly prescribed medication belongs to the penicillin class. (Courtesy Dr. George Hripcsak)

each new penicillin prescription for a penicillin allergy. In the Knowledge category the triggering event is that the newly prescribed medication belongs to the penicillin class. Of course, the way to determine that might well be EHR specific. Later in this MLM you see the logic to evaluate the decision and the action to take if its value is 'true'.

Keep in mind that each MLM can drive one decision so, in practice, a group of MLMs may be required. In Fig. 8.21 five MLMs help provide CDS for the use of warfarin, a common but potentially dangerous, blood thinner often given to patients after thrombotic strokes caused by clots or to treat other potential blood clotting problems. Too little of the drug may mean another stroke. Too much can cause

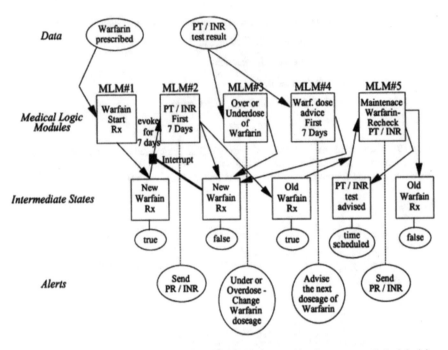

Fig. 8.21 An example of how a set of Arden MLMs can be combined to create a clinical decision support run. This example is for the appropriate prescribing of Warfarin, a notoriously difficult drug (for preventing blood clots) to manage. (Courtesy Dr. George Hripcsak)

excessive, and even fatal, bleeding. Warfarin interacts with many other drugs, and even the patient's diet, so it is a tricky drug to manage. Physicians use a test called PT/INR to assess the degree of blood thinning to manage the warfarin dose in each patient. The modules – using clinical data about the patient, including their PT/INR – can make warfarin dosage recommendations.

There is now an Arden XML schema (Fig. 8.22) that details all four categories. In addition to the two categories we saw earlier, each MLM also contains management information to help maintain a knowledge base of MLMs and links to other knowledge resources. Health personnel can create MLMs and implement them in any EHR or clinical information system that conforms to the Arden specification.

Figure 8.23 is the part of an MLM that helps with a radiologic study that involves injecting the patient with a contrast dye to highlight the functioning of parts of their internal organs. It specifically deals with a lab test – the creatinine level – that measures kidney function. It is important to know the patient's kidney status before giving the dye, because the kidneys remove the dye and it can cause kidney damage so it shouldn't be given to a patient with already compromised kidney function. As a result, Arden needs to know the creatinine level and, to get it, {'dam'="PDQRES2"}, which is highlighted in Fig. 8.23, must be interpreted by the EHR in a particular hospital to fetch that value from its proprietary database.

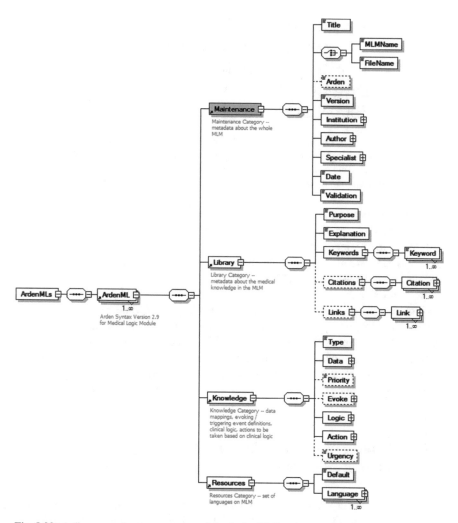

Fig. 8.22 A diagrammatic representation of the Arden XML schema detailing all four categories of Arden MLMs. (Courtesy Dr. George Hripcsak)

The purpose of Arden is to share CDS logic but, because EHRs have not been readily interoperable, to implement Arden locally each instance where the Arden MLMs need to access data in the hospital's EHR would have to be mapped to the data structure and representation of data items used by that EHR. Even the same EHR implemented in different hospitals may not do these things the same way due to decisions made when it was configured and installed. This understandably led to low Arden adoption. The impediment is so serious and widely known that it even has a name – the curly braces problem.

This leads us to one of the key advantages of FHIR and SMART on FHIR apps mentioned earlier as reasons for this discussion of CDS. FHIR-enabled EHRs would

```
creatinine := read {'dam'="PDQRES2"};
last_creat := read last {select "OBSRV_VALUE"
from "LCR" where qualifier in
("CREATININE",
"QUERY_OBSRV_ALL")};
```

Fig. 8.23 The part of an MLM that reads the last creatinine level from an EHR. The part of this in 'curly braces' must be replaced with the specific query needed to access data in the EHR where the CDS is being used. This need for site specific customization impeded Arden adoption and, as a result, came to be called the 'curly braces problem'. (Courtesy Dr. George Hripcsak)

present creatinine in a reasonably consistent form in a FHIR Observation Resource. As a result, a hypothetical CDS SMART on FHIR app need have no knowledge of the EHR's internal data structure or representation. It would not need to be adapted to each EHR, greatly reducing development cost and facilitating implementation and, hence, the likelihood of CDS adoption.

You may also recall that an impediment to MYCIN adoption, despite its superior performance to even infectious disease experts, was the time required to use an external tool that was not well integrated into the workflow and process of the busy physicians who could benefit from using it. FHIR and SMART on FHIR, if properly implemented, also resolve this long-standing impediment to CDS use. We will discuss this in the next chapter. First, we will look at some of the contemporary approaches to supporting CDS, including two that are employing FHIR.

8.14   Contemporary Tools for CDS Authoring and Dissemination

infobuttons Earlier we discussed an IOM report that identified seven 'information intensive' aspects of its vision that current EHRs often fail to provide. In that discussion we introduced the notion that EHRs would ideally be 'clinically adaptive' which is to say they could provide information relevant to the particular clinical context and the specifics of that patient.

infobuttons are an interesting attempt to provide this missing EHR functionality through context-specific links that would provide specific information of interest given factors such as the provider specialty, the clinical context, or the patient's characteristics. This was a decade after Arden began and was part of an effort to utilize the then new technologies of the Internet in clinical information systems.[17] infobutton development was led by James Cimino, MD now Director, Informatics Institute, School of Medicine, University of Alabama and Guilherme Del Fiol now

---

[17] https://www.ncbi.nlm.nih.gov/pubmed/10175348

Assistant Professor in the University of Utah's Department of Biomedical Informatics. It became an HL7 standard in 2010.[18]

The nuances of adding clinical context is explained in more detail in the following quotation from a 2012 article in the *Journal of Biomedical Informatics* that reported on a survey of 17 organizations that had attempted to implement the standard.

> infobuttons use the context of the interaction between a user and a clinical information system to predict the information needs that are most likely to occur and to retrieve content from online knowledge resources that may address these needs. Context can be represented in terms of a set of attributes that describe (1) the patient (e.g., gender, age); (2) the clinical information system user (e.g., discipline, specialty, preferred language); (3) the task being carried out in the clinical information system (e.g., order entry, problem list review, laboratory test result review); (4) the care setting (e.g., outpatient, inpatient, intensive care); and (5) the clinical concept of interest (e.g., a medication order, a laboratory test result, a problem).[19]

A revised specification provided an implementation guide for representing the context parameters that clinical information systems could send to knowledge resources via a URL. Special software called an Infobutton Manager would use those parameters to lookup the relevant information in a knowledge base and return a list of topics likely to be of interest. Each topic is, in turn, a link to a web page containing the matching information. As explained on the apparently inactive original infobuttons site "if a user is reviewing a patient's prothrombin time (a test of blood coagulation), an Infobutton Manager might provide links to various references about drugs that affect prothrombin time (such as warfarin sodium). If the patient is an adolescent or adult female, some of the links might be specifically related to pregnancy and breastfeeding recommendations. If the patient is a patient at New York Presbyterian Hospital, a link will be provided to the relevant age-specific hospital guidelines for the use of warfarin sodium."[20]

Curating content at this level of specificity is challenging so the National Library of Medicine and the University of Utah worked on a tool called Librarian Infobutton Tailoring Environment (LITE) aimed at supporting non-technical personnel (such as clinicians and librarians) to specify the resources that an Infobutton manager should provide, given a specific set of circumstances (for example, a nurse reviewing an adolescent female patient's potassium test result).

Today the effort is housed in the OpenInfobutton project that provides a current HL7 compliant release 2.2 version of the Infobutton Manager and LITE.[21]

MAGICapp MAGIC is a research driven initiative based in Denmark whose goal is to improve the creation, dissemination and maintenance of digital clinical practice guidelines, evidence summaries and decision support tools. It is led by Dr. Per

---

[18] http://www.hl7.org/implement/standards/product_brief.cfm?product_id=208

[19] https://www.sciencedirect.com/science/article/pii/S1532046411002206

[20] http://www.infobuttons.org/

[21] https://github.com/VHAINNOVATIONS/InfoButtons

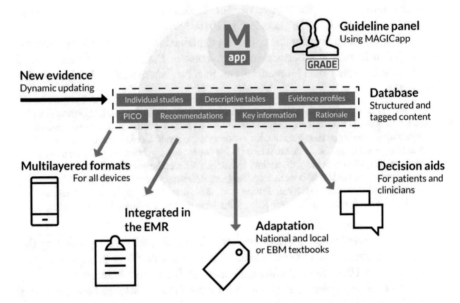

Fig. 8.24 The MAGICapp pipeline begins with new medical evidence which is put into a structured and tagged format suitable for the various purposes shown. (Courtesy Dr. Per Olav Vandvik)

Olav Vandvik a professor in the School of Medicine of the University of Oslo. It developed the MAGIC Authoring and Publication Platform (MAGICapp), a web-based platform for structured digital authoring, dissemination and dynamic updating of evidence summaries, recommendations and clinical decision support tools.

The MAGICapp authoring pipeline is shown in Fig. 8.24. It, in turn, rests on two separate concepts – PICO and GRADE. PICO is strategy for formulating clinical questions and search strategies. The acronym stands for its four elements which are:

Patient, **P**opulation or **P**roblem: The characteristics of the patient or population (demographics, risk factors, pre-existing conditions, etc.) or what is the condition or disease of interest?
Intervention: What is the intervention under consideration?
Comparison: What is the alternative to the intervention?
Outcome: What are the relevant outcomes?[22]

GRADE is a widely used methodology for grading the quality (or certainty) of medical evidence and the strength of recommendations derived from it.[23]
MAGICapp is well described by this paragraph from the MAGIC Project site:

---

[22] http://linkeddata.cochrane.org/pico-ontology

[23] http://www.gradeworkinggroup.org/

The PICO linked data model applied in MAGICapp breaks down the concept of a guideline document into discrete elements of content. This way of digitally structuring data allows content to be published in multilayered and flexible formats, usable on all devices and facilitates adaptation and dynamic updating of individual recommendations in a living guideline model. The linked data structure – combined with a coding module for annotating PICO questions and recommendations with terms from various structured terminologies and making the data available through an API – allows for connections (EHR, other platforms), re-use and sharing of data across the Ecosystem.[24]

Several interesting examples of MAGICapp are posted on the MAGIC Project site. You should try one or more since it is far easier to understand the tool by seeing it than it is to adequately explain it in text. One robust example compares two surgical treatments for severe aortic stenosis in low-intermediate risk patients.[25]

We will now look at two examples of contemporary clinical decision support using FHIR. These could have been placed later on but the seemed to fit topically here. They do refer, albeit lightly, to some concepts we haven't formally discussed yet so you may wish to read or re-read them after the two chapters on FHIR and SMART on FHIR that follow.

Case Study: Zynx Health

The company was founded in 1996 by a group of clinicians at Cedars-Sinai Medical Center in Los Angeles as a wholly owned subsidiary of the hospital. It was acquired by Hearst in 2005. An earlier pilot project had demonstrated the value of evidence-based guidelines in hospitals, suggesting to the founders that there was an opportunity to improve patient care.

The company's clinical experts accumulate and appraise peer-reviewed research, national guidelines, and performance measures. The evidence sources are classified and summarized with the goal of providing its clients with unbiased coverage of the most trustworthy research and quality measures. The evidence is offered as user-customizable products that are typically delivered through software as a service and include:

A system used by hospitals and outpatient physicians for developing and maintaining order sets based on clinical evidence, making use of rules, reminders and other tools to assist with physician decision making.

A care plan development system designed for hospital nursing staff and interdisciplinary teams, helping clinicians customize evidence-based plans of care.

A reference resource online database of clinical evidence drawn from medical and interdisciplinary literature, peer-reviewed research, and national guidelines and performance measures that is the content foundation on which other Zynx Health products are based.

(continued)

---

[24] http://magicproject.org/research-and-tools/living-guidelines/
[25] https://www.magicapp.org/app#/guideline/1308

A knowledge management solution that maximizes the value of your computerized provider order entry (CPOE) system and aligns your organization across desired clinical and financial goals.

As illustrated in Fig. 8.25, the Zynx Health Developer Program offers a FHIR API that makes over 800 PlanDefinition resources available containing evidence-based care guidance. Third-party solution developers can utilize these resources to assist providers in their delivery of patient care through adherence to these evidence-based standards.

Interested potential developers can sign up at the developer site and receive an API key that will make two complete PlanDefinition resources available for demonstration purposes.[26] Additionally, a demonstration application for using the API key can be downloaded from the Zynx Health API demonstration GitHub repository.[27]

The company has plans to make its evidence-based content available through a Zynx FHIR API into SMART on FHIR applications and CDS Hooks.

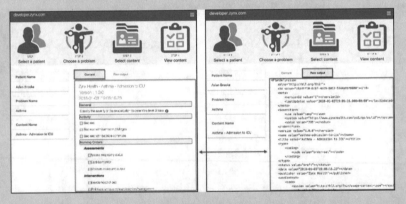

Fig. 8.25 On the right is an example of a Zynx Health care guideline is delivered as an XML formatted FHIR PlanDefinition resource. On the left is an illustration of how that resource might be used in a provider-facing app to help determine the appropriate level of care for a newly admitted asthma patient

Apervita Earlier we discussed an article in the Harvard Business Review that discussed the transformative impact that APIs have had in other industries and said that it "health care industry followed suit, the impact on the quality and cost of care, the patient's experience, and innovation could be enormous".[28]

---

[26] http://developer.zynx.com

[27] https://github.com/zynxhealth/api-demo

[28] https://hbr.org/2015/12/the-untapped-potential-of-health-care-apis

A great deal of the discussion around APIs in this book will center on apps that use the FHIR standard to access data in EHRs. However, a second transformative factor in healthcare and industry in general is the increasing use of analytics and the various forms of machine learning to derive knowledge from large datasets. Many clinical decision support tools would benefit greatly from or would depend entirely on their ability to access the most up-to-date analytic algorithms to inform and personalize the advice they give to providers or patients. Those analytic algorithms in turn could continuously learn and improve based on the information they glean from interacting with the widest possible user group. Rather than baking these analytic algorithms into software it would therefore be advantageous to make analytics a web service accessible by APIs.

That is the mission of Apervita. The company provides what it describes as a secure, self-service platform for the development and delivery of analytic tools which it says can scale to the extent that its users require. The company says it can integrates with many EMR systems to ingest data in real time or batch, or through the Apervita web service APIs. It can return the results of essentially any arbitrary clinical calculation such as an alert, score, text, quality report, chart, or dashboard within a browser window, iFrame, the recipient EMR's native alerting framework, or other capabilities supported by the EMR including SMART on FHIR.

Similar to the smartphone app marketplaces you are familiar with and the similar FHIR app galleries you will read about later, Apervita offers its clients a Marketplace to share or even market their analytic tools to others in the Apervita user community.

8.15   Recap

Standards provide more structured ways of exchanging, interpreting and ultimately using information. Since the early days of HL7 and the other standards development organizations we have discussed, healthcare standards have evolved and typically grown more complex over time. An important consideration in any standards development effort is not just how well the standard represents the nuances in the information being represented but the success of the standard in increasing the use of the information to help solve practical real world challenges. The discussions in this chapter provide the reader with important background and context from which the FHIR standard was born.

Chapter 9
FHIR

9.1 The Origins of FHIR

In 2011, Australian HL7 standards guru, Grahame Grieve, proposed a new interoperability approach he called "Resources for Health (RFH)". He said it would define a set of objects to represent granular clinical concepts for use on their own or aggregated into complex documents. As a result, it would be 'composable' – so that, unlike with complex C-CDA documents, developers could request only the information needed for their particular use case. In part, because of that, he said that this flexibility could offer "coherent solutions for a range of interoperability problems".

He went on to say that "Technically, RFH is designed for the web; the resources are based on simple XML, with an http-based RESTful protocol where each resource has [a] predictable URL. Where possible, open Internet standards are used for data representation."[1]

Earlier that year, the HL7 board had authorized a "Fresh Look" Task Force to examine, given current technologies, the best way to create interoperability solutions with no preconditions on what those solutions might be. The idea was not to start all over but rather, based on prior attempts by HL7 and other groups, the task force would consider new approaches to interoperability. Grahame joined the effort and, at its outset, posted several key questions on his blog: "What are we trying to do? What do our customers want? What exchanges are we trying to serve? Are we doing syntax or semantics? Does the market even want semantics? HL7 has two quite different stakeholders in vendors and large [healthcare] programs – can they agree on what they want?"[2]

As we discussed much earlier, Grahame ended up leading the resulting standards development effort, renamed Fast Healthcare Interoperability Resources (FHIR).

---

[1] http://www.healthintersections.com.au/?p=502
[2] http://www.healthintersections.com.au/?p=137

© Springer International Publishing AG, part of Springer Nature 2018
M. L. Braunstein, *Health Informatics on FHIR: How HL7's New API is Transforming Healthcare*, https://doi.org/10.1007/978-3-319-93414-3_9

We have touched on FHIR throughout the earlier parts of this book. Now we will look at it in more detail.

9.2 Grahame's Philosophy

As with any HL7 standards development effort, decisions are made by consensus but it's clear that Grahame's vision guides FHIR development so, before we delve into the standard, it is important to understand his philosophy.

FHIR begins with the concept of a simplified data model for healthcare. The FHIR data model consists of Resources, as Grahame proposed for RFH. Unlike HL7's C-CDA and RIM, resource contents are intentionally limited. FHIR's 80% rule is the operative guideline and it informally states that each resource should contain only those data elements agreed to by 80% or more of the participants in the development effort. Of course, as Grahame has posted, there are other considerations, such as safety and consistency.[3]

All standards efforts must grapple with the tension between comprehensiveness and usability. In essence, this approach means that, in FHIR development, usability is a top priority.

Grahame's second principle is around how to verify the standard. Historically, a vote of standards developers approves the standard they created. With FHIR, input from software developers who have worked to solve real world problems using proposed FHIR resources, APIs and other elements of the standard, guides further development. Again, usability is a priority.

The third principle is the use of cross-industry web technologies to search for resources and to Create, Read, Update and Delete them (CRUD). This makes the standard more approachable and, hence, more attractive to modern developers and provides the basis for FHIR-based mobile and web apps that can, at least in principle, be EHR agnostic.

9.3 FHIR Modules

The FHIR specification encompasses far more than direct patient care. Its scope is best appreciated by reviewing this list of its current division into these 13 modules:

Foundation: The basic definitional and exchange infrastructure on which the rest of the specification is built

Implementer Support: Services to help implementers make use of the specification

---

[3] http://www.healthintersections.com.au/?p=1924

Security & Privacy: Documentation and services to create and maintain security, integrity and privacy

Conformance: How to test conformance to the specification, and define implementation guides

Terminology: Use and support of terminologies and related artifacts

Linked Data: How to use RDF, and how ontologies are used when defining the FHIR content (more about this later)

Administration: Basic resources for tracking patients, practitioners, organizations, devices, substances, etc.

Clinical: Core clinical content such as problems, allergies, and the care process (care plans, referrals) + more

Medications: Medication management and immunization tracking

Diagnostics: Observations, Diagnostic reports and requests + related content

Workflow: Managing the process of care, and technical artifacts to do with obligation management

Financial: Billing and Claims support

Clinical Reasoning: Clinical Decision Support and Quality Measures

We will refer to some of the more interesting potential of some of these in what follows but those interested in exploring them in detail should refer to the FHIR Overview page which enumerates and links to discussions of each of them.[4]

9.4   FHIR Resources

As you now know, the FHIR data model is being instantiated in a set of what will probably be more than 150 modular "Resources". The current, Release 3 version of the standard, organizes its 147 resources and related documentation into the five levels in Fig. 9.1 (not all of the clinical resources are listed in this graphic). This same graphic is also organized by the modules listed above within these levels.

This graphic is useful for understanding the scope of the standard and its most important resources, particularly for patient care. However, it is a simplified summary of the standard. In what follows we will be referring to the complete FHIR Resource Index page that lists virtually all of the currently defined resources.[5]

At the time of this writing the levels and their resource counts on that page were:

Level 1: Foundation – The standard's basic framework (**30 resources**)

Level 2: Base – Support for implementation and binding to external specifications (**26 resources**)

Level 3: Clinical – Structural and process elements of real world healthcare systems (**37 resources**)

Level 4: Financial – Record-keeping and data exchange (**43 resources**)

---

[4] https://www.hl7.org/fhir/overview.html

[5] https://www.hl7.org/fhir/resourcelist.html

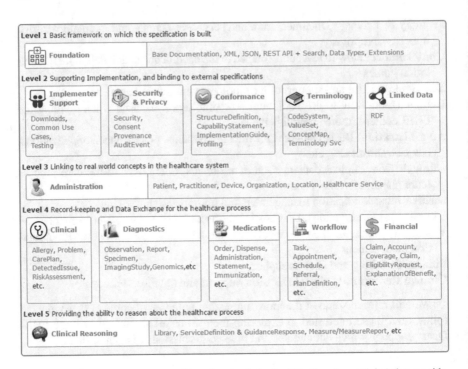

Fig. 9.1 FHIR resources are grouped into five levels (and within them by modules) that provide: (1) a basic framework; (2) support for implementation; (3) linkage to the healthcare system; (4) clinical documentation and data exchange; and (5) support for clinical reasoning. (Courtesy HL7)

Level 5: Specialized – Providing the ability to reason about healthcare processes (**11 resources**)

The levels are also illustrated in Fig. 9.2 which more functionally subcategorizes each level and adds a sixth level that consists of Profiles and Graphs, two tools that are used to specify the application of FHIR to a particular use case. We will discuss these later on in this chapter.

At present, FHIR is in its third version of a standard for trial use (STU3) and there are 147 resources on the Resource Index page, around the estimated number for the initial normative version of FHIR that will likely be released in late 2018. You may see the term Draft Standard for Trial Use (DSTU), but STU is now HL7's preferred term. An STU4 will precede the first normative version.

It is likely that the number of resources will increase over time and existing ones may change, so the HL7 FHIR Resource Index page will be the most up-to-date source of this information. It summarizes the resources and their sub-divisions in several useful ways. If you visit this page note that the number after each resource indicates its maturity on a scale of 0–5. Those numbers will increase over time as standards definition proceeds. At present, the resources in the lower numbered of the five levels listed earlier are generally more mature.

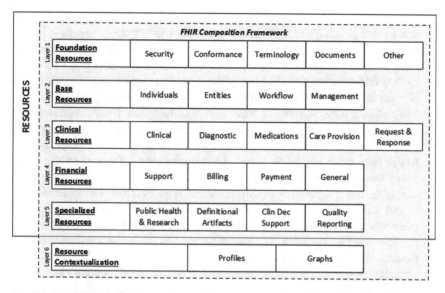

Fig. 9.2 The FHIR health information model sub-categories organized into layers based on how frequently that are anticipated to be used. The categories at the top layers are the most commonly used so they contain the FHIR resources that support the largest number of healthcare transactions and therefore need to be the most consistently defined and tightly governed. (Courtesy HL7)

There are a few particularly interesting resources that support the FHIR framework or suggest interesting new directions for the standard:

- The definition of a resource is itself contained in the Resource resource that is not listed on the Resource Index page but can be found in the Foundation Module. This is only one of 17 resources used in this module but, unlike the others, it is strictly definitional so there would never be an actual Resource resource in a patient record.
- Earlier, we said that Value Sets can be represented in a resource. It is the ValueSet resource in the Terminology Module that "provides an overview and guide to the FHIR resources, operations, coded data types and externally-defined standard and FHIR-defined terminologies that are used for representing and communicating coded, structured data in the FHIR core specification and profiles".
- Later, we will discuss FHIR profiles. These are represented in a computable StructureDefinition resource found in the Conformance division of the Foundation level on the Resource Index page.[6]
- The Group resource in the Individuals divisions of the Base level on the resource list page specifies a particular set of patients, providers, medications, etc. for purposes such as reporting or tracking.[7] They are likely to be an important component of the newly proposed but not yet formally defined Bulk Data Protocol we will discuss later.

---

[6] https://www.hl7.org/fhir/structuredefinition.html
[7] https://www.hl7.org/fhir/group.html

- The current FHIR information retrieval paradigm is 'pull'. An app or other system or tool needing information sends a RESTful API query to obtain it. The Subscription resource in the 'Other' division at the Foundational level is used to create a 'push' of FHIR resources that match criteria specified in the resource. Updated information would be sent along without being further requested. This is a potentially important direction for the standard. Helping clinicians keep track of their patients could be facilitated if updated information automatically appears and is brought to their attention. One important example of this are critical laboratory results which can have an impact on timely and appropriate care or can help make timely decisions about discharging patients from the emergency department or hospital bed.[8]

- One of the major limitations of current clinical systems is their lack of knowledge of and inability to support optimal workflow and process. This not only leads to inefficiency but medical errors such as when a critical laboratory result doesn't reach its intended recipient. The Workflow Module (not a module in the sense of the ones we just discussed) also suggests an important new direction for the standard that Grahame specifically mentioned in an interview I did with him in the fall of 2017. The primary aspects of this 'informative' group of resources are:

  - How do we ask for another person, device or system to do something?
  - How do we track the linkages and dependencies between activities – actions to their authorizations, complex activities to individual steps, protocols to plans to orders, etc.?
  - How do we define what activities are possible and the expected order and dependencies of the steps within those activities? i.e. process/orchestration definition

- Finally, Level 5 includes support for Clinical Reasoning. Earlier, we discussed clinical decision support and the benefit of tight EHR integration using FHIR as a key strategy to obtain provider adoption of these tools. The new CDS Hooks specification, developed by the same group that created SMART on FHIR, supports the integration of clinical logic into EHRs to, among other things, automatically indicate the availability of applicable FHIR apps when a patient's clinical situation suggests they might be of value. A goal of FHIR STU 4 is to unify the CDS Hooks specification with the Clinical Reasoning module. This should be another important step toward broader use of CDS.

9.5   FHIR Resource Representations

There are examples on the HL7 FHIR site of most Resources in XML, JSON and Turtle formats. XML (eXtensible Markup Language) and JSON (Javascript object notation) are two text-based standards for open data sharing on the Internet. XML

---

[8] https://www.hl7.org/fhir/subscription.html

is older but JSON is being widely adopted because it is less verbose and, hence, easier to parse. You saw examples of both when we discussed HL7 V2 and V3 messages. The JSON representation of FHIR resources is by far the most commonly used at present.

The list of FHIR modules included "Linked Data" and its brief description referred to RDF. Turtle (Terse RDF Triple Language) is a format used by the World Wide Web Consortium (W3C) in its Resource Description Framework (RDF) data model for the proposed semantic web in which data in web pages is structured and tagged so that it can be read (some might say 'understood') directly by computers. To support this RDF extends the linking structure of the Web using Universal Resource Identifies (URIs) to describe relationships between things as well as the two ends of the link in a "triple" comprising a subject, predicate and object, each of which can be a URI. This should remind you of our earlier discussion of semantic interoperability.

Figure 9.3 suggests how this construct could power a semantic web that understands entities and their relationships.[9] In it, as is often the case, the Mona Lisa is at the center of attention. Bob is a person who likes this painting created by Leonardo da Vinci. As in our earlier discussion of semantic interoperability, it is worth considering whether there is a need for a semantic formalism in an era when machines are increasingly figuring these things out. There is a discussion of representing FHIR Resources in RDF on the HL7 site.[10] There is also a W3C Semantic Web Health

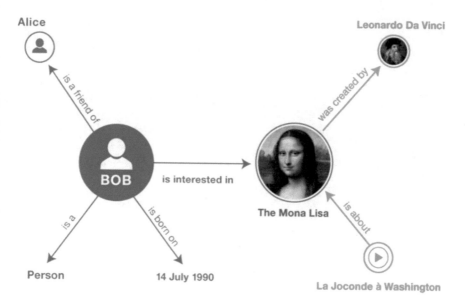

Fig. 9.3 A simple graph illustrates the kind of relationships the RDF seeks to codify. (Courtesy W3)

---

[9] https://www.w3.org/TR/rdf11-primer/

[10] https://www.hl7.org/fhir/rdf.html

Care and Life Sciences Interest Group.[11] The possible use of this representation would be of obvious interest to the semantic interoperability community and an HL7 group has been created to further define the use of RDF in FHIR.[12]

9.6 FHIR Resource Examples

Figure 9.4 illustrates some key parts of an XML-formatted FHIR patient resource:

- Local information, such as its last update and its local resource ID (the value that can be used to find this instance of the resource on this particular server)
- An HTML presentation of the key information for human consumption or for display by the receiving system if it is not able to interpret the structured information

```
<Patient xmlns="http://hl7.org/fhir">
  <id value="glossy"/>
  <meta>
    <lastUpdated value="2014-11-13T11:41:00+11:00"/>
  </meta>
  <text>
    <status value="generated"/>
    <div xmlns="http://www.w3.org/1999/xhtml">
      <p>Henry Levin the 7th</p>
      <p>MRN: 123456. Male, 24-Sept 1932</p>
    </div>
  </text>
  <extension url="http://example.org/StructureDefinition/trials">
    <valueCode value="renal"/>
  </extension>
  <identifier>
    <use value="usual"/>
    <type>
      <coding>
        <system value="http://hl7.org/fhir/v2/0203"/>
        <code value="MR"/>
      </coding>
    </type>
    <system value="http://www.goodhealth.org/identifiers/mrn"/>
    <value value="123456"/>
  </identifier>
  <active value="true"/>
  <name>
    <family value="Levin"/>
    <given value="Henry"/>
    <suffix value="The 7th"/>
  </name>
  <gender value="male"/>
  <birthDate value="1932-09-24"/>
  <careProvider>
    <reference value="Organization/2"/>
    <display value="Good Health Clinic"/>
  </careProvider>
</Patient>
```

Resource Identity & Metadata

Human Readable Summary

Extension with URL to definition

Standard Data:
- MRN
- Name
- Gender
- Birth Date
- Provider

Fig. 9.4 The key components of a FHIR resource include local data, human readable data, any extensions used in this resource and its contents as defined by the FHIR specification. (Courtesy HL7)

---

[11] https://www.w3.org/2011/09/HCLSIGCharter

[12] http://wiki.hl7.org/index.php?title=RDF_for_Semantic_Interoperability

```
{
  "resourceType": "AllergyIntolerance",
  "id": "medication",
  "text": {
    "status": "generated",
  },
  "clinicalStatus": "active",
  "verificationStatus": "unconfirmed",
  "category": [
    "medication"
  ],
  "criticality": "high",
  "code": {
    "coding": [
      {
        "system": "http://www.nlm.nih.gov/research/umls/rxnorm",
        "code": "7980",
        "display": "Penicillin G"
      }
    ]
  },
  "patient": {
    "reference": "Patient/example"
  },
  "assertedDate": "2010-03-01",
  "recorder": {
    "reference": "Practitioner/13"
  },
  "reaction": [
    {
      "manifestation": [
        {
          "coding": [
            {
              "system": "http://snomed.info/sct",
              "code": "247472004",
              "display": "Hives"
            }
          ]
        }
      ]
    }
  ]
}
```

Fig. 9.5 The structured part of a JSON-formatted FHIR AllergyIntolerance Resource documents that it is active but unconfirmed, relates to a medication and is deemed critical. It goes on to specify that the medication is Penicillin G (coded in RxNorm); to indicate who the patient and provider are by providing their resource IDs on this server; and to document that the allergy is manifested by hives (coded in SNOMED-CT). (Courtesy HL7)

• Any local extensions to the standard
• The standard defined data content.

Figure 9.5 is an example of the structured part of a JSON-formatted FHIR AllergyIntolerance Resource. It begins by showing that this allergy problem is clinically 'active' but is 'unconfirmed'. Patients often report allergies, but that does not

Code	Display	Definition
food	Food	Any substance consumed to provide nutritional support for the body.
medication	Medication	Substances administered to achieve a physiological effect.
environment	Environment	Any substances that are encountered in the environment, including any substance not already classified as food, medication, or biologic.
biologic	Biologic	A preparation that is synthesized from living organisms or their products, especially a human or animal protein, such as a hormone or antitoxin, that is used as a diagnostic, preventive, or therapeutic agent. Examples of biologic medications include: vaccines; allergenic extracts, which are used for both diagnosis and treatment (for example, allergy shots); gene therapies; cellular therapies. There are other biologic products, such as tissues, that are not typically associated with allergies.

Fig. 9.6   The Value Set or list of allowed values for the allergy category. (Courtesy HL7)

mean they experienced an actual allergy as opposed to a medication side effect. This allergy is for Penicillin G and, according to the Mayo Clinic, research has shown that penicillin allergies may be over diagnosed and that patients may report a penicillin allergy that has never been confirmed. A misdiagnosed penicillin allergy may result in the use of less appropriate or more expensive and possibly more dangerous antibiotics.

Next it documents that this allergy is in the Medication Category. 'Medication' is one of four allowed values for an allergy/intolerance category. Figure 9.6 shows this Value Set – the list of allowed values for this data element – taken from the relevant page on the HL7 FHIR site.[13] Food, environmental and biologic allergies (such as to an X-ray contrast agent or a vaccination) can also be documented using this resource. Knowing that it is a medication allergy could be helpful to software so that it knows to check on it when prescribing new medications.

Next is a reference to an RxNorm code or RXCUI – 7980 – for Penicillin G. Actually, penicillin is a group of antibiotics and this is only one part of that group. Patients allergic to one member of the group are presumed to be allergic to the entire group, so an RXCUI of 70618 might have been a better choice. You can verify that and find out what this RXCUI is for using the RxNorm tool suggested to you earlier.

---

[13] http://hl7.org/fhir/ValueSet/allergy-intolerance-category

Path	Definition	Type
AllergyIntolerance.clinicalStatus	The clinical status of the allergy or intolerance.	Required
AllergyIntolerance.verificationStatus	Assertion about certainty associated with a propensity, or potential risk, of a reaction to the identified substance.	Required
AllergyIntolerance.type	Identification of the underlying physiological mechanism for a Reaction Risk.	Required
AllergyIntolerance.category	Category of an identified substance.	Required
AllergyIntolerance.criticality	Estimate of the potential clinical harm, or seriousness, of a reaction to an identified substance.	Required
AllergyIntolerance.code	Type of the substance/product, allergy or intolerance condition, or negation/exclusion codes for reporting no known allergies.	Example
AllergyIntolerance.reaction.substance	Codes defining the type of the substance (including pharmaceutical products).	Example
AllergyIntolerance.reaction.manifestation	Clinical symptoms and/or signs that are observed or associated with an Adverse Reaction Event.	Example
AllergyIntolerance.reaction.severity	Clinical assessment of the severity of a reaction event as a whole, potentially considering multiple different manifestations.	Required
AllergyIntolerance.reaction.exposureRoute	A coded concept describing the route or physiological path of administration of a therapeutic agent into or onto the body of a subject.	Example

Fig. 9.7 The list of references to value sets and external codes (terminology bindings) used in the AllergyIntolerance Resource

Finally, the resource documents that the allergy manifests itself as hives, an outbreak of swollen, pale red bumps or plaques (or wheals) on the skin. The SNOMED-CT code provided should be for hives. You can verify that with the SNOMED browser you may have used earlier.

The references to RxNorm and SNOMED-CT illustrate an important characteristic of FHIR we mentioned earlier – it builds upon existing standards – via "Terminology Bindings" that specify these relationships. Figure 9.7 shows all the terminology bindings for this one resource. There is a page on the FHIR site listing all the terminology bindings referenced in the standard.[14]

---

[14] http://hl7.org/fhir/terminologies-valuesets.html

Level	Code	Display	Definition
1	A	Annulled	Marriage contract has been declared null and to not have existed
1	D	Divorced	Marriage contract has been declared dissolved and inactive
1	I	Interlocutory	Subject to an Interlocutory Decree.
1	L	Legally Separated	
1	M	Married	A current marriage contract is active
1	P	Polygamous	More than 1 current spouse
1	S	Never Married	No marriage contract has ever been entered
1	T	Domestic partner	Person declares that a domestic partner relationship exists.
1	U	unmarried	Currently not in a marriage contract.
1	W	Widowed	The spouse has died

Fig. 9.8 The value set for marital status used in FHIR derives from HL7 RIM

Figure 9.8 illustrates how FHIR also rests on key elements from earlier HL7 standards. For example, the marital status value set comes from HL7 V3. If you go to the Detailed Description tab on the Patient Resource page on the FHIR site, you will find Patient.maritalStatus.[15] Click on its Marital Status Codes link and you will get to this table.

9.7   FHIR Resource Activity

David Hay, a leader in the FHIR community and the developer of the excellent clinfhir web tool for exploring FHIR has posted a video that demonstrates the use of clinfhir to create FHIR Resources.[16] This is something you may wish to try yourself using David's wonderful tool.

9.8   FHIR Extensions

The 80% rule inevitably means that the FHIR standard does not cover all use cases. FHIR extensions serve the needs of a use case with specific requirements or a user community with specific interests or needs. HL7 maintains a registry of extensions

---

[15] https://www.hl7.org/fhir/patient.html

[16] https://videos.files.wordpress.com/qIAx7Xbo/clinfhir_demo_dvd.mp4

Fig. 9.9 An extension to the FHIR Allergy Intolerance resource documents the age at which it resolved and might be particularly useful in a pediatric environment. (Courtesy HL7)

so that implementers can take advantage of already defined extensions and avoid duplication.[17]

Figure 9.9 is a simple example of a FHIR extension to the AllergyIntolerance Resource we saw earlier. It provides a structured way of recording the age at which the allergy resolved. Childhood allergies to food do often resolve during later childhood or adolescence and would be a common use case for this extension. Note that to assure consistency, the status of a resource using this extension should be 'resolved'.

9.9   FHIR Resource IDs

Earlier in Figs. 9.4 and 9.5 you saw examples of FHIR Resource IDs. FHIR Resources have a logical identity or ID that is unique among all resources of the same type on each FHIR server. Just to be clear, this ID will usually be different on each FHIR server so it cannot, for example, establish patient identity across servers. However, there is a discussion of the use of FHIR for a Master Patient Index on the Patient resource page[18] and there is even the value set shown in Fig. 9.10 to indicate the quality of a patient query match. This illustrates how rich the FHIR standard is becoming. You should use the many public web resources and tools to explore it further.

The unique identifier of a resource instance is an absolute URI constructed from the server base address, the resource type and the logical ID. When the literal identity is an HTTP address, this address can generally retrieve or manipulate the resource as illustrated in Fig. 9.11 using a public instance of the HAPI FHIR server. In this example the URI is **https://fhirtest.uhn.ca/baseDstu3/Patient?identifier=3535**. The three parts of the URL are: the server base address of **fhirtest.uhn.ca/baseDstu2/**; the resource type of **patient**; and the resource ID of **3535**. In the body of the figure you see that, as it should, the internal reference to the Resource ID matches the one in the URL.

---

[17] https://www.hl7.org/fhir/extensibility-registry.html

[18] https://www.hl7.org/fhir/patient.html

Code	Display	Definition
certain	Certain Match	This record meets the matching criteria to be automatically considered as a full match.
probable	Probable Match	This record is a close match, but not a certain match. Additional review (e.g. by a human) may be required before using this as a match.
possible	Possible Match	This record may be a matching one. Additional review (e.g. by a human) SHOULD be performed before using this as a match.
certainly-not	Certainly Not a Match	This record is known not to be a match. Note that usually non-matching records are not returned, but in some cases records previously or likely considered as a match may specifically be negated by the matching engine

Fig. 9.10 The value set for use in indicating how precisely a patient has been matched to information presumably provided by an outside system. (Courtesy HL7)

Fig. 9.11 The use of a unique identifier to create an absolute URI to retrieve that resource. In this case it is https://fhirtest.uhn.ca/baseDstu3/Patient?identifier=3535. (Courtesy University Health Network)

You should easily be able to do something similar using any public FHIR server. We will discuss those later. Keep in mind that they are essentially 'playgrounds' and there is no assurance that data will persist so do not be surprised if the examples given in the book don't work for you. Instead, simply search for a Resource giving no specifications to bring up all the instances of that resource. However, you should also definitely study the dropdown lists of search criteria available to you. Use one to search for something fairly likely to be present (e.g. patients with a Family Name of "Smith"). Pick one and then note its literal URL and the three parts that make it up. Before moving on note the use of "DSTU" in the URL. As mentioned earlier,

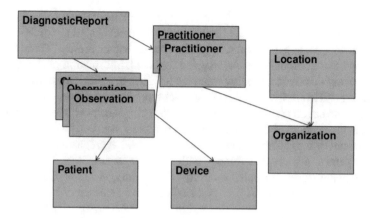

Fig. 9.12 FHIR can be thought of as creating a network of information via the references within many FHIR resources to other related resource. These references are of obvious value to app developers. (Courtesy HL7)

you are likely to run into this even though the preferred term is now STU. They are equivalent.

Resource IDs can be up to 64 characters long and contain any combination of upper and lowercase ASCII letters, numerals, "-" and ".". Note that the ID for the patient earlier was "example" and that could actually be an ID.

The AllergyIntolerance resource example (Fig. 9.5) we have been exploring has ID references for the patient ('example') who has this allergy as well as to the practitioner ('13') who recorded it. As shown in Fig. 9.12, using these references FHIR Resources create a network of information that is somewhat like the hierarchy of relationships coded into SNOMED-CT.

9.10 FHIR Enabling Existing Systems

In real-world scenarios health system vendors usually create FHIR Resources from data stored in a proprietary clinical database designed and built before FHIR existed. When a FHIR app or REST API asks for a resource special software, often called a FHIR adapter or server, must know how to get the needed data elements from the system's database and package them into the FHIR Resource format.

The data that populates that FIHR resource will therefore depend on the underlying system. If that system stores clinical observations only as text then, even though FHIR includes them, LOINC codes will not be available in the FHIR resources that system creates. Of course, the designers of the FHIR adapter might make the effort to add any missing codes but, at present, that is not typical.

Moreover, there are often many LOINC codes for similar lab tests or clinical observations so different systems may provide different codes. Commercial products are starting to appear that seek to aggregate and even normalize data from different sources to create a more consistent set of FHIR resources. Earlier we saw that Human API and the Google Health Cloud have developed a middleware layer to add missing codes and potentially validate codes that are present. These functions have long been a part of software designed for health information exchanges so it is not surprising that these are being extended to FHIR by HIEs. One early example is InterSystems' HealthShare HIE software that typically aggregates data from multiple electronic health records and other systems using whatever interfaces are available from the vendors of those systems. The company says that "now all the data for an individual or a population can be transformed and aggregated as a FHIR representation, and applications developed with FHIR can access it."[19]

Many major healthcare enterprise software vendors, including Epic, Cerner and MEDITECH (the three most widely installed) have released at least preliminary versions of a FHIR server. As of this writing Epic and Cerner and some other EHR vendors have released integrated support for the SMART on FHIR app platform and MEDITECH indicates that it plans to release support (but, at present, this may be limited by some vendors to patient facing apps because that's a requirement for Meaningful Use Stage 3). We will look at examples of this later on. Finally, again at present, most vendor support is only for reading resources. Support for the rest of CRUD is more difficult for vendors to implement because of the need to implement various checks and edits and other safeguards necessary to maintain the integrity and consistency of their databases so support beyond read only will appear more slowly.

HL7 maintains a web page that summarizes the ever-growing list of health IT system vendors and other organizations interested in FHIR along with a link to their sites, contact information and a brief description of their intended use of the standard.[20]

9.11   FHIR API

This discussion is a bit more technical but it should be possible for all readers to get the general concepts.

Client-server is an architecture in which data is stored on a central system (server) and is made accessible to remote (client) computers or other devices via some network. The Internet is, of course, a dramatic global extension of this architecture. Some protocol or standard is required so that the clients and servers understand each other. Representational State Transfer (REST) is by far the most common protocol on the Internet and it relies on a stateless, client-server, cacheable communications protocol that is usually HTTP. You use this protocol on a daily basis when you access public and even private web sites.

---

[19] https://www.intersystems.com/fhir/
[20] http://wiki.hl7.org/index.php?title=Organizations_interested_in_FHIR

Statelessness essentially means that all of the necessary information to handle the request is contained within the request itself, whether as part of the URI, query-string parameters, body, or headers. You saw an example of that earlier when we retrieved patient 3535. The FHIR server only needed the information in the URL to do that. Thus, any machine in a vast server farm (such as those managed by Amazon, Google and others) can handle the request, return the appropriate response and immediately be available for the next request. A moment's reflection should reveal to you why this is an attractive characteristic for anyone managing a large public site. The response from the server to the client is similarly self-contained.

In FHIR, the four HTTP verbs POST, GET, PUT, and DELETE implement CRUD. POST creates new resources. GET reads existing resources. PUT and DELETE update or remove existing resources. I mention these because you may well see them if you further explore FHIR.

Figure 9.13 is a complete list of the currently defined logical FHIR interactions. This part of the standard is at Maturity level 5 so it is essentially in final form. Note its three sections: Instance, Type and Whole System interactions.

The Instance section deals with individual resources. Note that PATCH, a fifth HTTP verb, can also be used when a client is seeking to minimize its bandwidth utilization or, in the more likely scenario, when a client has only partial access or support for a resource, a fairly likely situation in healthcare given the emphasis on privacy and security.

Instance Level Interactions

read	Read the current state of the resource
vread	Read the state of a specific version of the resource
update	Update an existing resource by its id (or create it if it is new)
patch	Update an existing resource by posting a set of changes to it
delete	Delete a resource
history	Retrieve the change history for a particular resource

Type Level Interactions

create	Create a new resource with a server assigned id
search	Search the resource type based on some filter criteria
history	Retrieve the change history for a particular resource type

Whole System Interactions

capabilities	Get a capability statement for the system
batch/transaction	Update, create or delete a set of resources in a single interaction
history	Retrieve the change history for all resources
search	Search across all resource types based on some filter criteria

Fig. 9.13 A complete list of the defined FHIR API interactions between a FHIR client/app and a FHIR server. They are divided into Instance, Type and Whole System interactions. (Courtesy HL7)

The Type section deals with a group of related resources. Note that here FHIR defines HISTORY and SEARCH, two interactions of particular interest in healthcare.

HISTORY retrieves a set (called a "bundle") of either a particular resource, all resources of a given type, or all resources supported by the system. It can be quite useful given the clinical importance of trends and changes in patient status.

Earlier you should have inspected the variety of search criteria available in FHIR. This should have convinced you that SEARCH is a particularly rich and important part of the FHIR standard. In the simplest case, a search is executed using a GET operation where a series of name=[value] pairs encoded in the URL specify the parameters.

Searches are constrained to one of three contexts in order to control what resources are considered:

- A specified resource type: **GET [base]/[type]?parameter(s)**
- A specified compartment that might be a specific patient, as illustrated in Fig. 9.14, perhaps with a specified resource type in that compartment: **GET [base]/Patient/[id]/[type]?parameter(s)**
- All resource types: **GET [base]?parameter(s)** where the parameters must be common to all resource types

Finally, in the Whole System section, the Capabilities operation retrieves the server's Capability Statement describing the server's current operational functionality. Clients connecting to a FHIR server can use the capabilities interaction to check whether they are version and/or feature compatible with the server or whether it can support a specific use case of interest. Along with the StructureDefinition resource we discussed earlier as the computable representation of a FHIR Profile, this is a key part of the FHIR conformance layer that consists of the StructureDefinition and Conformance resources as well as an Implementation Guide.

If the reasons for this are not clear to you, keep in mind that the FHIR specification describes a set of resources and, at present, three frameworks for exchanging resources between different systems. Because of its general nature and wide applicability, the rules made in this specification are generally fairly loose so different

Fig. 9.14 An example of a FHIR API to retrieve a specific patient using their ID on the server where the query is performed. (Courtesy University Health Network)

applications may not be able to interoperate because of how they use optional features.

Applications claim conformance to one (or more) of these three FHIR exchange frameworks (or paradigms in which FHIR may be used):

RESTful FHIR: The RESTful API for resource-by-resource interaction between a FHIR app or client and a FHIR server.

FHIR messaging: Message based exchange that involves a group (bundle) of FHIR resources that together could represent something like an HL7 message.

FHIR documents: Document based exchange involves a typically larger bundle of FHIR resources that together could represent something like an HL7 C-CDA document.

Importantly to developers the FHIR resources are consistent across these three paradigms further simplifying app development.

An important future direction for the FHIR API is signaled by the following statement in ONC's Draft Trusted Exchange Framework released in January 2018: "Within twelve (12) months after the FHIR standard with respect to Population Level Query/Pulls has been formally approved by HL7, each Qualified HIN [HIE] shall cause its Broker to be able to initiate and respond to all Query/Pulls for as many individuals as may be requested by another Qualified HIN in a single Query/Pull."[21] Later we will discuss the proposed FHIR Bulk Data Protocol to respond to this requirement.

9.12 FHIR Profiles and Graphs

The FHIR specification provides a "platform" that creates a common foundation on which a variety of different solutions can be implemented. Often this general specification requires adaptation to particular use cases. FHIR Profiles provide a means of constraining the standard for a specific use case. FHIR Profiles can specify a variety of restrictions such as which of the values specified in the standard are allowed within a value set for the use case(s) implemented using the profile. Alternatively, they might specify another value set that is more extensive or used in a special context. They can require that a data element be present. They can also add extensions to FHIR.

According to the FHIR site "Profiles are used to extend, constrain, or otherwise contextualize resources for a given purpose. Graphs are compositions of resources, or webs of resource[s], that contain attributes of their own."[22] Importantly Resources, Profiles and Graphs are all represented in their own resources making the specification of FHIR for a use case itself computable. This greatly facilitates the develop-

---

[21] https://www.healthit.gov/sites/default/files/draft-trusted-exchange-framework.pdf

[22] https://www.hl7.org/fhir/overview-arch.html

ment of FHIR apps and tools that are correctly using the standard as intended for their particular use case(s).

Implementation Guides specify Profiles and contain the Conformance Resource specification for that use case. Argonaut is the best-known example of an Implementation Guide. The Argonaut Project is a collaboration of many private, major EHR vendors and other organizations to promote industry adoption of modern, open interoperability standards to meet the requirements of Meaningful Use and related regulations. This is a complex topic best explored using the information posted on the HL7 FHIR site.[23]

The FHIR site also provides a more comprehensive discussion of the FHIR architectural framework and how it aligns with some widely used standard frameworks (such as the TOGAF Architecture Development Method, the Zachman Framework and HL7's Services Aware Interoperability Framework (SAIF)).[24]

9.13   FHIRPath

Arguably one of the most attractive features of FHIR for developers is its composability. Each FHIR resource is an independent object and can be retrieved using an appropriate API query. This is distinctly unlike earlier standards such as C-CDA where an entire clinical document must be retrieved even if only one specific part of that document (such as a patient's problem list) is of interest. To explore this in some detail we will use a new tool called FHIRPath that is implemented on David Hay's wonderful clinfhir web tool.[25]

Among other things, FHIRPath further extends composability to individual data types or even their values. FHIRPath is a path-based navigation and extraction language, somewhat like the XPath language defined by W3 and which can be used to navigate through elements and attributes in an XML document. FHIRPath does the same for FHIR and can navigate through and retrieve specific sub-elements of a large FHIR Bundle which might, for example, be an entire patient record. When the FHIR Bulk Data Protocol is implemented this bundle could even be all the records of patients in a "group" that could, for example, specify only patients with a particular insurance plan.

The technical details of FHIRPath are beyond the scope of this book but we can easily see how useful it could be using some examples. One of clinfhir's many attractive features is the generation of some sample data so that a new patient begins with a reasonable facsimile of a medical record. This is illustrated in Fig. 9.15 where we have checked the 'Generate samples' box.

The resulting 72 FHIR Resources are indexed of the left in Fig. 9.16. The single Patient resource is selected and its contents are displayed on the right in a human

---

[23] http://www.fhir.org/guides/argonaut/r2/

[24] https://www.hl7.org/fhir/overview-arch.html#framework

[25] http://clinfhir.com/

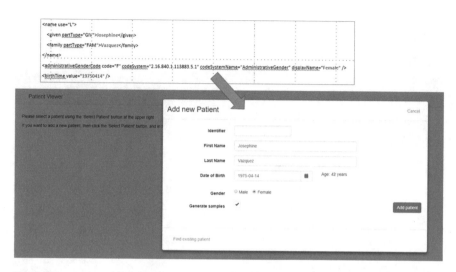

Fig. 9.15 An example of using clinfhir to generate a new patient with a variety of FHIR resources containing make believe data by checking the 'Generate Samples' box. (Courtesy David Hay)

Fig. 9.16 On the left clinfhir indexes the FHIR resources generated for the make believe patient requested previously. The single Patient resource is selected and is displayed on the right in a human friendly tree view. (Courtesy David Hay)

friendly 'tree view'. Note that this patient's Patient Resource has been assigned the ID 299014, a number that can be used to easily retrieve it or, depending on the capabilities of an individual FHIR server, to specify that other resources, such as medications or clinical observations, are to be retrieved, but only for this patient.

The Patient Resource can be represented in XML or in JSON, as shown in Fig. 9.17. It should be easy to see that the data that is shown in the tree view is the same as in this JSON representation.

Tree   Text   Json   XML   References

Patient/299014   Download

```json
{
  "resourceType": "Patient",
  "id": "299014",
  "meta": {
    "versionId": "1",
    "lastUpdated": "2017-11-01T10:19:15.452-04:00"
  },
  "text": {
    "status": "generated",
    "div": "<div xmlns=\"http://www.w3.org/1999/xhtml\">Josephine Vazquez</div>"
  },
  "name": [
    {
      "use": "official",
      "text": "Josephine Vazquez",
      "family": "Vazquez",
      "given": [
        "Josephine"
      ]
    }
  ],
  "gender": "female",
  "birthDate": "1975-04-14"
}
```

Fig. 9.17 The JSON representation of the patient confirms the source of the data shown in the tree view. (Courtesy David Hay)

clinfhir's FHIRPath exploration arm makes it easy to create a bundle containing our patient's entire record, as shown in the middle part of Fig. 9.18.

It may be a simplistic representation of a patient record but, displayed in Microsoft Word, this bundle occupies 96 pages and is 5401 words long. We already know that we can use the FHIR API to retrieve a part of this bundle such as only the Condition resources. Suppose that we aren't interested in the entire contents of the Condition resources because we only want to display the patient's medical problems as text in a hypothetical FHIR app. That information is contained in those Condition resources and it would certainly be possible to parse it from them. FHIRPath makes that simple to do, as shown in the right pane of Fig. 9.17. There, the FHIRPath expression

```
Bundle.entry.where(resource.resourceType='Condition').resource.
code.text
```

produces exactly the text needed to display the patient's medical problems as you can see on the right below this expression.

To give another example, the following FHIRPath expression uses the appropriate LOINC code to produce a list of measurements for a particular clinical observa-

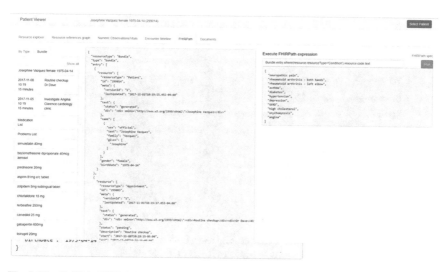

Fig. 9.18 clinfhir's FHIRPath tool creates a bundle containing our patient's entire record. This is ideal for further exploration using the FHIRPath language. (Courtesy David Hay)

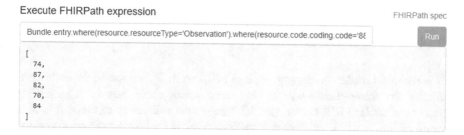

Fig. 9.19 This FHIRPath expression retrieves only the numeric values of the clinical observations with a LOINC code of 8867-4. (Courtesy David Hay)

tion, as shown in Fig. 9.19. I leave it to you to use the LOINC code to figure out what specific clinical measurement this is.

```
Bundle.entry.where(resource.resourceType='Observation').
where(resource.code.coding.code='8867-4').resource.valueQuantity.
value
```

Capabilities such as FHIRPath make app creation far easier and faster leading to more rapid innovation and more usable software tools.

For example, from this list it should be easy to create a graph over time, as shown in Fig. 9.20. While this may seem to be a simplistic example many commercially available EHRs cannot yet graph results like this and a FHIR app using FHIRPath would both be quite simple to implement and could be a useful tool for providers using one of those EHRs.

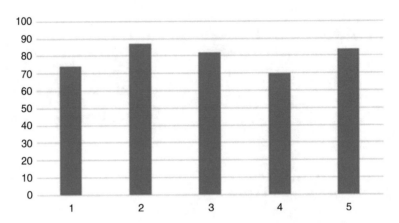

Fig. 9.20 The results of the FHIRPath expression shown in Fig. 9.18 are graphed to illustrate how approachable working with clinical data becomes if it is FHIR accessible

Later we will discuss the role that FHIRPath will play in the FHIR specifications for quality measurement and clinical decision support.

9.14   Public FHIR Servers

Non-technical readers wishing to explore FHIR on their own would do well to use clinfhir. For those readers with at least some coding skills there is an increasing number of public FHIR servers and HL7 maintains a directory of these. It is well worth reviewing the descriptive information posted in the directory to get a sense of the variety of services available.[26]

Many of these public servers will share a user interface because they are based on the open source HAPI implementation of the FHIR specification done in Java by James Agnew, a co-chair of HL7 and an editor of the FHIR standard.[27] He is also technical lead for SMILE CDR which provides commercial products based on the code. Among other things SMILE CDR provides a FHIR enabled HIE solution[28] Readers with an interest can download the HAPI code from GitHub and create their own FHIR server.[29] The most widely known public test server based on HAPI is provided by the University Health Network in Toronto.[30]

---

[26] http://wiki.hl7.org/index.php?title=Publicly_Available_FHIR_Servers_for_testing

[27] http://hapifhir.io/

[28] https://smilecdr.com/

[29] https://github.com/jamesagnew/hapi-fhir

[30] http://fhirtest.uhn.ca/

Commercial EHR vendors are creating 'sandboxes' to facilitate development of FHIR apps for their platform. Among the first of these are the major EHR vendors Cerner,[31] Epic[32] and Allscripts®.[33]

The data available on most publicly accessible FHIR servers can be of highly variable quality, utility and persistence. One interesting exception is SyntheticMass developed by the MITRE Corporation, a not-for-profit company that operates multiple federally funded research and development centers and has had a long-standing involvement in health informatics. The data stored on this server consists of what MITRE calls "realistic but fictional residents of the state of Massachusetts" that "statistically mirror the real population in terms of demographics, disease burden, vaccinations, medical visits, and social determinants." The data was produced by MITRE's open source Synthea tool.[34] The entire dataset can be graphically explored on the site and it can also be accessed via FHIR queries.[35] A number of example queries are provided on the site.[36]

9.15   FHIR Recap

As time passes the momentum toward FHIR seems to be growing at an increasing rate. To get a sense of this it is well worth attending one of HL7's FHIR Application Roundtable meetings that provide a large number of short presentations/demos from a wide variety of sources including providers, vendors (including increasingly large companies), academic institutions, start-ups and even individuals. HL7 members can view prior presentations on the organization's site.[37] Anyone attending this series of meetings would likely note the increasing size of the commercial companies presenting and the increasing scope and apparent maturity of the products, systems and platforms they are offering.

It now seems clear that HL7's fresh look may indeed have led us to the long-sought solution to interoperability, or at least its first two layers – transport and structure. In what follows we will explore what can be done using FHIR to help solve the many problems of healthcare delivery with which we began this book. First, we will explore SMART on FHIR, the increasingly ubiquitous and also long sought universal health app platform.

---

[31] http://fhir.cerner.com/

[32] https://open.epic.com/AppExchange/Sandbox

[33] https://tw171.open.allscripts.com/FHIR/

[34] https://synthetichealth.github.io/synthea/

[35] https://syntheticmass.mitre.org/fhir/

[36] https://syntheticmass.mitre.org/api.html

[37] https://www.hl7.org/events/fhirapps.cfm

Chapter 10
SMART on FHIR

10.1 A Grand Challenge

A universal health app platform to support informatics-based innovations in care delivery, no matter what the underlying EMR, was a long-held dream of the academic health informatics community. In 2010, the ONC awarded $15 million to the Boston Children's Hospital Computational Health Informatics Program and the Harvard Medical School Department of Biomedical Informatics to create and initially develop just such an app platform.

The result was SMART, an acronym for "Substitutable Medical Apps, Reusable Technology" that was intended to be an open, standards-based technology with a robust community of app developers. The SMART effort developed an API that leveraged web standards such as HTML, JavaScript, OAuth, Resource Description Framework (RDF) and widely adopted data standards such as RxNorm, LOINC and SNOMED-CT in order to present 'predictable data payloads' thus obviating the need for apps to deal with the details of the underlying health information system's data structures and representations. Systems that implemented all or part of the SMART API were called 'containers'.

This approach placed a burden on that underlying system. "To enable such substitutability, the SMART architecture imposes a substantial burden of normalization on any container, allowing apps to know upfront what data to expect and what each data element means. An important implication is that some containers may need to reshape underlying data to support the SMART API." To avoid adding to this burden the SMART API was read only.[1]

Despite this, according to an article by the SMART team "through mid-2013, EHR vendors were not receptive. At the time, third-party medical apps were not yet an important near-term business driver. In addition, vendors raised objections to the SMART data model and API because they had not been developed in conjunction

---

[1] https://dash.harvard.edu/bitstream/handle/1/10436330/3384120.pdf?sequence=1

© Springer International Publishing AG, part of Springer Nature 2018
M. L. Braunstein, *Health Informatics on FHIR: How HL7's New API is Transforming Healthcare*, https://doi.org/10.1007/978-3-319-93414-3_10

with the health IT standards community. Vendors also identified specific technical obstacles to adoption, such as our focus on clinical data to the practical exclusion of other types, e.g., administrative data; no perceived advantage to RDF over "plain" extensible markup language (XML) or JavaScript Object Notation (JSON); concerns about our solution to a write API; and the lack of population-level API."[2]

10.2   The Evolution to SMART on FHIR

As the preceding section documents, SMART development preceded FHIR but the approach had difficulty gaining traction with the major EHR vendors in part because it had not been developed within a standard setting body. As HL7's FHIR effort became more significant, the SMART project shifted its strategy to the development of a technology layer that builds on the FHIR API and resource definitions. Today it is known as SMART on FHIR but the project team often uses SMART to refer to it. SMART on FHIR is now widely adopted by the same major EHR vendors that objected to the original SMART technology.

10.3   SMART on FHIR Technology Stack

As we explained earlier, FHIR provides a set of "core" data models (resources) but many FHIR fields and value sets may not be constrained to support specific requirements across varied regions and use cases. To enable substitutable health apps as well as third-party application services, SMART on FHIR applies a set of "profiles" that provide developers with expectations about the vocabularies they should use to represent medications, problems, labs, and other clinical data.

In the United States SMART on FHIR has adopted the profiles outlined in the Argonaut Implementation Guide we discussed in the previous chapter.[3] This takes advantage of the broad adoption of Argonaut by the EHR vendor community. Also as part of the Argonaut Project, the five largest EHR vendors either have or will build SMART on FHIR into current or future releases of their products. The vendors, the SMART on FHIR team and HL7 are working together to standardize the SMART on FHIR API in HL7 specifications.

Clearly, third party apps must not access protected health information without establishing trust in who is using those apps and respecting patient privacy choices. To facilitate this, SMART on FHIR provides login and data access authorization models based on the OpenID Connect[4] and OAuth2[5] standards that are already

---

[2] https://academic.oup.com/jamia/article/23/5/899/2379865?rss=1

[3] http://www.fhir.org/guides/argonaut/r2/

[4] http://openid.net/connect/

[5] https://oauth.net/2/

widely used on the Internet and elsewhere. In what follows we will look at how SMART on FHIR uses these technologies.

Through the SMART on FHIR Genomics and SMART CDS Hooks efforts, the SMART on FHIR project is helping define the next generation of FHIR based standards for the clinical use of genomic data and the integration of clinical decision support into provider workflows. We will also discuss these efforts later.

10.4   Developer Support

To support running apps within the EHR's user interface, SMART on FHIR allows web apps to run inside browser widgets or inline frames. It also supports native and mobile apps.

To make it easier for developers to get started building apps the SMART on FHIR project offers a set of open source libraries for HTML5/JavaScript,[6] Apple's iOS[7] and Python.[8]

SMART on FHIR also offers a free, web-based API "sandbox" that developers can use to test their apps as well as a locally installable version that developers can download and run on their own system. Figure 10.1 shows the public sandbox that supports installing an app and testing it using sample patients, practitioners and data. For example, there is an extensive set of vital signs data that is available for

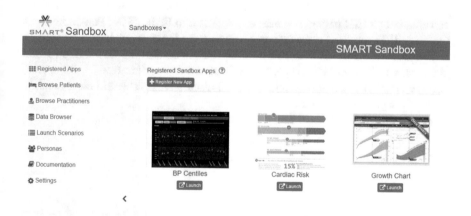

Fig. 10.1 The SMART on FHIR public sandbox supports installing an app and testing it against sample patients, practitioners and data. (Courtesy SMART)

---

[6] http://docs.smarthealthit.org/clients/javascript/

[7] http://docs.smarthealthit.org/Swift-SMART/

[8] http://docs.smarthealthit.org/clients/python/

[9] http://docs.smarthealthit.org/sandbox/

testing. Developers can also create patient and practitioner 'personas' and use them
in "Launch Scenarios" to test their app.[9]

Projects often have specific data requirements so SMART offers FRED, an open
web-based, interactive FHIR resource editor to help developers create sample data
for their apps.[10] You may wish to try using FRED to build a FHIR Resource of your
choice. As shown in Fig. 10.2, you can input some elements of the data elements
contained in a FHIR patient resource and export it in FHIR JSON Patient resource
format. FRED code and more information about the tool is available on GitHub.[11]

SMART's C3-PRO is a software library that integrates the SMART on FHIR
platform with ResearchKit, Apple's open source framework to enable iOS apps to
become tools for medical research.[12] We will discuss ResearchKit in more detail in
the final chapter of this book.

Finally, the SMART on FHIR site offers extensive documentation.

10.5 OAuth2

OAuth2 is a widely used authorization framework that enables web and desktop
applications, as well as mobile devices, to obtain specified, constrained access to
user accounts on an HTTP web service.

OAuth2 defines four roles:

---

[10] http://docs.smarthealthit.org/fred/

[11] https://github.com/smart-on-fhir/fred

[12] https://github.com/C3-PRO/c3-pro-ios-framework/blob/master/C3PRO.podspec

- Resource owner
- Resource server
- Authorization server (can be the same server as the Resource server)
- Client (in the case of SMART this would be a FHIR app)

The resource server hosts the protected information. The authorization server verifies the identity of the prospective user of that information and then issues access tokens to the client application. You might recall that in the discussion of Apple Health's support of FHIR for patients to access their records it was Geisinger's OAuth2 authorization server that managed access to that data.

An OAuth2 scope specifies the level of access (such as 'read' or 'write') that the application is requesting. In the case of SMART on FHIR the scope might specify what FHIR Resources can be accessed for a specified patient and how much of CRUD the app can perform on those resources.

In SMART on FHIR when an EHR user launches an app, it gets a "launch request" notification. The app asks for the permissions it needs using OAuth scopes such as **patient/*.read**.

In the case of a launch within an EHR charting session, the scope "Patient/*. read" refers to the patient whose chart is in use at the time of the launch. The term for this is "in-context" and this request only makes sense within a charting session. The app does not yet know who that patient is when it requests the scope. It learns that from the access token response, where the "patient" property will provide the FHIR Patient Resource ID.

To review, in response to its scope request, the app receives an access token with the permissions it needs – including access to clinical data and context like:

- which patient is in-context in the EHR
- which encounter is in-context in the EHR
- the physical location of the EHR user

These so called 'launch parameters' are an extremely important capability of SMART on FHIR. As we discussed earlier, tools such as clinical decision support have existed for decades but have not found wide use. To coin a small pun, providing this context literally allows the app to be 'smarter' avoiding work for its clinical user and thereby increasing the attractiveness of using the app.

10.6 Scopes and Permissions

In the example we just discussed our hypothetical scope request contained a wildcard (*). Through it, the client app is seeking permission to obtain all available *and future* FHIR resources stored on the server for the current patient. This raises the issue of patient privacy. The authorization server asking a patient to authorize a SMART app requesting **patient/*.read** should inform the patient that they are being asked to allow the app access to all currently available and future FHIR resources that might contain potentially more sensitive data such as their genomic variations. Of course, as we discussed when we looked at HealthVault, giving the patient usable tools to make an informed decision about what they are willing to share remains a challenge.

However, the granular nature of FHIR Resources could certainly facilitate the development of such a tool. This would be particularly true if, as illustrated in Fig. 10.3, which is a MedicationRequest example from the FHIR site, the FHIR resources contained the reason (e.g. condition) for which orders and procedures were ordered. If they did, then a patient might specify that they wish to share only information related to a particular clinical problem and the app could figure out what that information is.

At present, these 'cause and effect' relationships are not specified in all the potentially applicable resources. Even if they were, this is not just an issue for the standard. A solution would require vendor support of the idea in their EMR and provider willingness to take the time to indicate these relationships. Alternately, this may well be another opportunity for machine learning to infer relationships from the data recorded over many patient encounters. To give a simple example, if virtually all patients with diabetes have had a hemoglobin A1c test and almost no other patients have had the test done then it would be possible to infer that this test is ordered for diabetes and hemoglobin A1c results could be automatically released to an app to which the patient has given permission to obtain all information about their diabetes.

The authorization server may not grant the level of access requested by the app. For example, an app with the goal of obtaining read and write access to our old friend, the patient's AllergyIntolerance resources, requests the clinical scope of **patient/AllergyIntolerance.***. The authorization server may respond in a variety of ways as shown in Fig. 10.4. For example, the authorization server might grant only read access as illustrated in the third granted scope in the figure. App designers should anticipate this and their apps should consider the permissions granted to them and they may need to change their behavior accordingly.

10.7   OpenID Connect

As we just discussed, in many cases SMART apps launch within the context of an existing EHR charting session where the provider, patient and encounter are clear. In other contexts, SMART may need to use OpenID Connect, a simple identity layer on top of the OAuth 2.0 protocol. It enables apps to verify the identity of the end user based on the authentication performed by the authorization server and obtain basic profile information about the end user.

To use OpenID Connect the app asks for the OpenID scope when it requests authorization and it will have access to a UserInfo endpoint that exposes user information referred to in OpenID as "claims". In SMART these include the end user's name and their unique National Provider Identifier (NPI), a 10-digit number

```
          "resourceType": "Medication",
          "id": "med0320",
          "code": {
            "coding": [
              {
                "system": "http://snomed.info/sct",
                "code": "324252006",
                "display": "Azithromycin 250mg capsule (product)"
              }
            ]
          }
        }
      ],
      "identifier": [
        {
          "use": "official",
          "system": "http://www.bmc.nl/portal/prescriptions",
          "value": "12345689"
        }
      ],
      "status": "active",
      "intent": "order",
      "medicationReference": {
        "reference": "#med0320"
      },
      "subject": {
        "reference": "Patient/pat1",
        "display": "Donald Duck"
      },
      "context": {
        "reference": "Encounter/f001",
        "display": "encounter who leads to this prescription"
      },
      "authoredOn": "2015-01-15",
      "requester": {
        "agent": {
          "reference": "Practitioner/f007",
          "display": "Patrick Pump"
        },
        "onBehalfOf": {
          "reference": "Organization/f002"
        }
      },
      "reasonCode": [
        {
          "coding": [
            {
              "system": "http://snomed.info/sct",
              "code": "11840006",
              "display": "Traveller's Diarrhea (disorder)"
            }
          ]
        }
      ]
    }
```

Fig. 10.3 This MedicationRequest resource documents that the drug (Azithromycin) is being given for a specific reason (traveler's diarrhea) facilitating patient permissions for sharing data about a specific condition while not sharing other data the patient might feel is sensitive. (Courtesy HL7)

Granted Scope	Notes
patient/AllergyIntolerance.*	The client was granted exactly what it requested: patient-level read and write access to allergies via the same requested wildcard scope.
patient/AllergyIntolerance.read patient/AllergyIntolerance.write	The client was granted exactly what it requested: patient-level read and write access to allergies. However, note that this was communicated via two explicit scopes rather than a single wildcard scope.
patient/AllergyIntolerance.read	The client was granted just patient-level read access to allergies.
patient/AllergyIntolerance.write	The client was granted just patient-level write access to allergies.

Fig. 10.4 The authorization server may respond in a number of ways to the scope request of "patient/AllergyIntolerance.*" asking for access to read and write AllergyIntolerance resources. (Courtesy SMART)

assigned by CMS to providers that uniquely identifies them across healthcare organizations.

10.8   SMART App User and Access Authorization

A SMART app must be registered with an EHR's authorization service before it can be run within that EHR and access data from it. The registration technology is up to the EHR vendor although SMART recommends the OAuth 2.0 Dynamic Client Registration Protocol. No matter how an app registers, at registration time it must register one or more fixed, fully specified URLs and redirect URLs so the EHR knows how to launch it.

EHRs can offer SMART apps to the provider in ways that greatly increase their likelihood of use. Figure 10.5 is an example of a SMART on FHIR app launching within Cerner's PowerChart® EHR. Note along the bottom left of the screen that the SMART on FHIR apps are in the same main menu that the physician uses for other charting activities. This greatly facilitates access to the apps and saves time by eliminating extra clicks. In this case, the widely used Meducation app has received the necessary scopes and permissions to obtain the patient's medications so no further input from the physician is required. Also, note that the display of the app's patient education information conveniently appears in the usual clinical data entry/presentation area of the EHR screen so, for all practical purposes, although its user interface is different, it appears to be functionally a part of the EHR.

Alternatively, as shown in Fig. 10.6, an app can launch as a standalone iOS or Android app. This is a design prototype done by a group of Georgia Tech students for an app that supports epilepsy patients via their smartphones. This kind of standalone patient facing app and apps for providers launching within an EHR session are probably the most common use cases for SMART on FHIR.

In an EHR launch, SMART passes an opaque[13] handle or pointer to the EHR context to the app as part of the launch URL. The app later will include this context

---

[13] One that does not directly expose any data.

Fig. 10.5 The Meducation app is launched from the main menu of the Cerner PowerChart EHR (note the highlighted menu choice) and its information is displayed in the same area where physicians normally do charting. The app has received the necessary scopes and permissions to obtain the patient's medications so no data input by the physician is required. (Courtesy Meducation)

handle as an OAuth authorization request parameter when it asks for authorization to access resources. Note again that the complete URLs of all apps approved for use by users of this EHR must be registered with the EHR's authorization server. When an app is launched standalone from outside an EHR context, it requests authorization to access a FHIR resource by redirecting its authorization request to the EHR's authorization server. Based on pre-defined rules and possibly end user authorization, the EHR authorization server either grants the request by returning an authorization code to the app's redirect URL, or denies the request. The app then exchanges the authorization code for an access token, which the app presents to the EHR's resource server to obtain the FHIR resource(s). If a refresh token comes along with the access token, the app may use this to request a new access token, with the same scope, once the original access token expires.

This has been just a basic discussion of SMART App Authorization. The important details are readily available on the excellent SMART on FHIR site.[14] There are also numerous web tutorials about OAuth2 and OpenID Connect. Google provides an interactive OAuth2 "playground".[15]

---

[14] http://docs.smarthealthit.org/authorization/

[15] https://developers.google.com/oauthplayground/

Fig. 10.6 A stand-alone
smartphone app developed
by a team of Georgia Tech
students supports patients
in managing their epilepsy

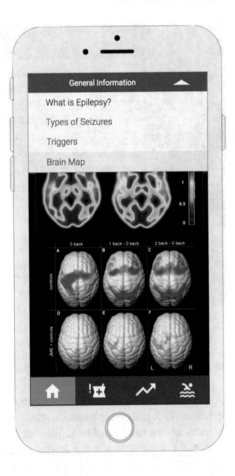

10.9   SMART Backend Services

A new arm of SMART supports developers of back-end services to use SMART to access FHIR resources by requesting access tokens from OAuth 2.0 compliant authorization servers. As with apps, the backend service must be authorized in advance and must:

- Run automatically, without user interaction
- Be able to protect a private key
- Require access to a population of patients rather than a single patient

Such a service might, for example, review incoming laboratory results, and generate clinical alerts when specific trigger conditions exist or it might maintain an external database of patient data for research purposes.

10.10 CDS Hooks

If you think back to our discussion of FHIR apps integrated into EHRs, it should be clear that the decision to initiate the apps rests with the EHR user. This, in turn, rests on the user recognizing that an app exists that might be of benefit in a particular patient care scenario. Clinicians are busy so it is easy to imagine that this might not happen. Very much along the lines first explored by Arden, the CDS Hooks initiative within SMART provides a means of embedding clinical logic within the EHR to invoke decision support, which might or might not be a full FHIR app. Unlike with Arden this occurs within a clinician's EHR workflow and, because of FHIR, there is no 'curly braces problem' since clinical data will be provided in a predictable, standard format.

A "hook" is functionality provided by software for users of that software to have their own code called under certain circumstances. Clinical examples of such circumstances might include:

- opening a new patient record
- authoring a new prescription
- viewing pending orders for approval

Under such circumstances CDS Hooks can display 'cards' to provide useful information or offer apps to help make optimal clinical decisions.

There are three kinds of CDS Hooks cards. **Information cards** convey text. **Alternative suggestion cards** provide an action different from the one contemplated. **App link cards** suggest the use of an app or reference materials.

Figure 10.7 is a screen shot from the interactive CDS Hooks demonstration tool to show how a hook inserted into the code for recording new prescriptions causes the display of a Suggestion Card. This example explains the potential savings if a generic equivalent replaces the brand name drug the provider input.[16]

This example raises an issue that may only slowly be addressed by EHR vendors. In this prescribing example it would obviously contribute to the acceptance of the advice if the physician could simply click on the 'change to generic' button at the lower left of the card. In practice this means that the EHR must allow 'write' access to its medication records. At present virtually all EHR vendor support of FHIR is read only. This is, of course, a regulatory requirement but writing to the EHR isn't. This will likely change over time but it may not occur as rapidly as many would like.

10.11 Specifying Quality Measures and Clinical Logic for Decision Support

The specification of clinical quality measures and clinical logic for decision support are somewhat related challenges. Quality measures must specify the population that is subject to the measure and the computations needed to evaluate the measure.

---

[16] http://cds-hooks.org

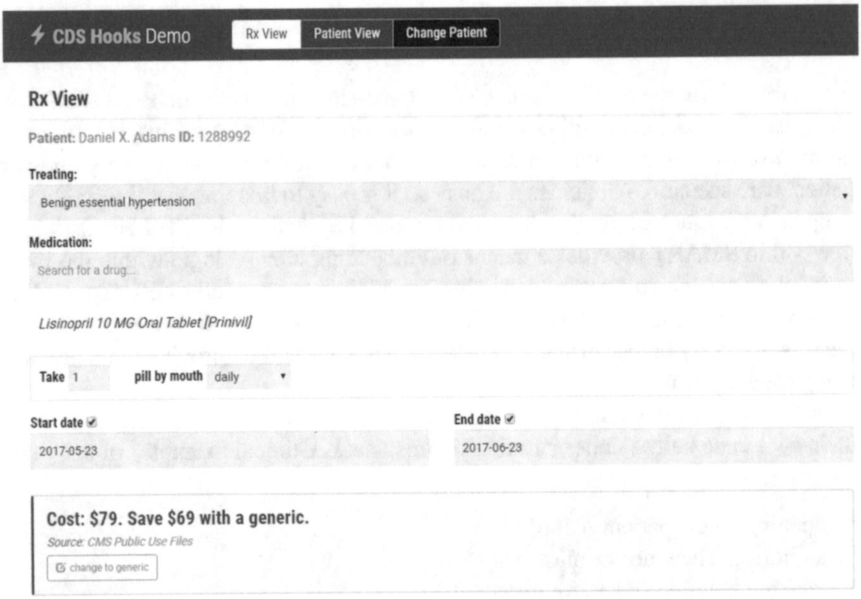

Fig. 10.7 Using FHIR, a CDS Hooks card realizes that the physician is prescribing a brand name drug for a patient's hypertension and displays this information in an attempt to suggest that the prescription might be for a lower cost generic medication saving $69. (Courtesy SMART)

Similarly, decision support artifacts must provide criteria describing which patients should be the recipient of a particular intervention.

For example, if the quality measure is the percentage of patients with diabetes who have had a hemoglobin A1C test (as specified by a list of LOINC codes) within the prior year (the numerator in the calculation), the applicable group of patients (the denominator in the calculation) would be those with diabetes (as specified by a list of condition codes in SNOMED-CT and/or ICD). Additional criteria such as age or gender might further stratify the population into subgroups. A decision support artifact might provide the criteria (again using the same list of condition codes in SNOMED-CT or ICD) to identify those patients who have diabetes but have not had the A1C test (using the same list of LOINC codes) done in the prior 12 months so that the clinician can be reminded to order the test.

Although standards exist for these purposes, the domains of quality measurement and clinical decision support have used different standards. Harmonization of these different approaches would enable broader sharing of computable clinical knowledge, as well as reduce the burden on authors and implementers responsible for producing and consuming that knowledge.

Earlier we discussed FHIRPath as a means or extracting specific data from one or more FHIR resources. Clinical Quality Language (CQL) is an HL7 standard used to describe the logic used by knowledge artifacts such as clinical decision support rules, quality measure logic, and conditions and actions under which specific sets of orders or care protocols would be administered to a patient. CQL expressions generally provide a means of computing new information from existing data. CQL is being extended to use FHIRPath as its core path language to represent paths within queries.[17]

For example, given a patient with multiple names, each of which has multiple givens, the following CQL query will return a single list containing all the given names:

```
expand (Patient.name X where X.given is not null return (X.given Y
where Y is not null))
```

The following FHIRPath expression will do the same thing:

```
{"resourceType": "Patient", "name": [{"given": ["John"]}]}
```

You can see an online demonstration of this FHIRPath expression.[18] CDS Hooks does not specify how the clinical logic to be embedded in the EHR will be specified but CQL with FHIRPath is at present the likely candidate for this role.

10.12   FHIR Genomics

Personalized or precision medicine are synonymous terms for the concept of using rich and potentially broad datasets about an individual patient to guide optimal diagnosis and treatment decisions that are as specific to that patient as possible. These datasets might supplement the traditional clinical record with behavioral, environmental, socioeconomic and genomic information. Behavioral data is largely the domain of mHealth, a later topic.

Genomic datasets are very large. Current estimates of a single patient's genome are around 3,200 MB or 3.2×10^9 base pairs. As a result, genomic data is unwieldy for direct use in patient care. The goal of the FHIR Genomics effort is to make genomic data usable and useful for patient care. Among other things, it is defining a new FHIR sequence resource, shown in part in Fig. 10.8, which contains the most clinically relevant data.[19] SMART on FHIR apps that can present that data to clinicians in useful ways.

Given the complexity of genomic data, it should not be surprising that this is perhaps the most complex FHIR resource yet specified. The specification includes instructions, as shown in Fig. 10.9, that explain how to document single nucleotide polymorphisms (SNPs), the most common type of genetic variation among people. SNPs occur approximately once in every 300 nucleotides, which means there are roughly 10 million SNPs in your genome. Their significance with respect to health and disease can vary greatly.

---

[17] https://github.com/cqframework/clinical_quality_language/wiki/FHIRPath:CQL::XPath:XQuery
[18] http://niquola.github.io/fhirpath-demo/#/
[19] https://www.hl7.org/fhir/sequence.html

10.8 Sequence Resource

10.8.1 Structure Diagram

Name	Flags	Card.	Type	Description & Constraints
Sequence	Σ I		DomainResource	Information about a biological sequence + Only 0 and 1 are valid for coordinateSystem Elements defined in Ancestors: id, meta, implicitRules, language, text, contained, extension, modifierExtension
identifier	Σ	0..*	Identifier	Unique ID for this particular sequence. This is a FHIR-defined id
type	Σ	0..1	code	aa | dna | rna sequenceType (Example)
coordinateSystem	Σ	1..1	integer	Base number of coordinate system (0 for 0-based numbering or coordinates, inclusive start, exclusive end, 1 for 1-based numbering, inclusive start, inclusive end)
patient	Σ	0..1	Reference(Patient)	Who and/or what this is about
specimen	Σ	0..1	Reference(Specimen)	Specimen used for sequencing
device	Σ	0..1	Reference(Device)	The method for sequencing
performer	Σ	0..1	Reference(Organization)	Who should be responsible for test result
quantity	Σ	0..1	Quantity	The number of copies of the seqeunce of interest. (RNASeq)
referenceSeq	Σ I	0..1	BackboneElement	A sequence used as reference + Only +1 and -1 are valid for strand + GenomeBuild and chromosome must be both contained if either one of them is contained + Have and only have one of the following elements in referenceSeq : 1. genomeBuild ; 2 referenceSeqId; 3. referenceSeqPointer; 4. referenceSeqString;
chromosome	Σ	0..1	CodeableConcept	Chromosome containing genetic finding chromosome-human (Example)
genomeBuild	Σ	0..1	string	The Genome Build used for reference, following GRCh build versions e.g. 'GRCh 37'
referenceSeqId	Σ	0..1	CodeableConcept	Reference identifier ENSEMBL (Example)
referenceSeqPointer	Σ	0..1	Reference(Sequence)	A Pointer to another Sequence entity as reference sequence
referenceSeqString	Σ	0..1	string	A string to represent reference sequence
strand	Σ	0..1	integer	Directionality of DNA (+1/-1)
windowStart	Σ	1..1	integer	Start position of the window on the reference sequence
windowEnd	Σ	1..1	integer	End position of the window on the reference sequence
variant	Σ	0..*	BackboneElement	Variant in sequence
start	Σ	0..1	integer	Start position of the variant on the reference sequence
end	Σ	0..1	integer	End position of the variant on the reference sequence
observedAllele	Σ	0..1	string	Allele that was observed
referenceAllele	Σ	0..1	string	Allele in the reference sequence
cigar	Σ	0..1	string	Extended CIGAR string for aligning the sequence with reference bases
variantPointer	Σ	0..1	Reference(Observation)	Pointer to observed variant information
observedSeq	Σ	0..1	string	Sequence that was observed

Fig. 10.8 Unsurprisingly given the complexity of the genome, the Genome Sequence FHIR resource, is itself very complex and this is only a part of it. (http://wiki.hl7.org/index.php?title=FHIR_ Genomics). (Courtesy HL7)

Each SNP represents a difference in a single DNA building block, called a nucle-otide. In this example the nucleotide thymine (T) in the observed sequence replaces the nucleotide guanine (G) in the reference sequence. This substitution is docu-mented in the FHIR Sequence resource.

Apps could access FHIR genomic resources from EHRs, genomic data reposito-ries or specialized sequencing systems. The key effort for a data provider is imple-mentation of a SMART on FHIR Genomics data adaptor, which creates a binding to convert between the standard SMART on FHIR Genomics format and the provider's native format.

KRAS is an oncogene that, when mutated, has the potential to cause normal cells to become cancerous. Figure 10.10 is an early example of Vanderbilt School of Medicine's prototype SMART Precision Cancer Medicine app usefully presenting genomic information in real-time by comparing a cancer patient's diagnosis-specific

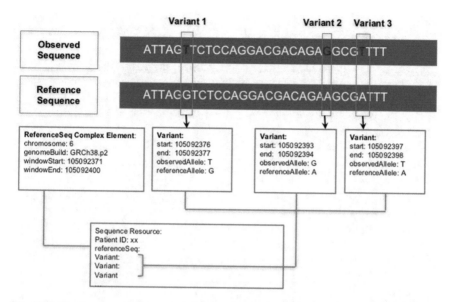

Fig. 10.9 The Sequence Resource specification includes instructions that explain how to document single nucleotide polymorphisms (SNPs), the most common type of genetic variation among people. Here the nucleotide thymine (T) in the observed sequence replaces the nucleotide guanine (G) in the reference sequence. (Courtesy HL7)

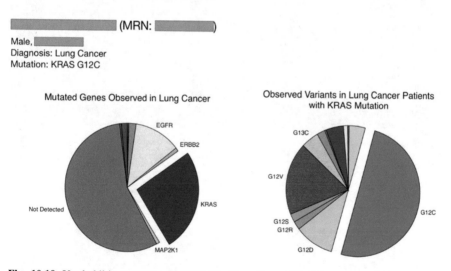

Fig. 10.10 Vanderbilt's prototype SMART Precision Cancer Medicine app usefully presents genomic information in real-time by comparing a cancer patient's diagnosis-specific detected gene mutations to a comparable population of cancer patients. (Courtesy SMART)

detected gene mutations to a comparable population of cancer patients. It should be clear to you how the Sequence Resource we just discussed would support an app such as this one.[20]

---

[20] https://apps.smarthealthit.org/app/smart-precision-cancer-medicine

Demonstration FHIR App: Timely Display of Renal Function in the ICU
Earlier in our discussion of the Arden Syntax we reviewed an example of a set of Medical Logic Modules (MLMs) that deal with the use of the creatinine level as a measure of kidney function in preparation for a patient having a dye study that could be dangerous if that function is compromised.

Kidney status is a clinical factor in many situations and can be particularly important for the very sick patients in an intensive care unit (ICU). What we didn't discuss earlier is that the creatinine level is a trailing indicator of kidney function. Creatinine is a byproduct of muscle metabolism that is removed from the blood by the kidneys. Blood circulates throughout the entire body so that process isn't instantaneous. There is an inevitable delay between a change in kidney function and the resulting change in creatinine level. In normal medical practice this isn't of any real significance because the delays are measured in hours. However, in a critical care situation it is important to know when kidney function changes including when it starts to improve so that treatment can be adjusted accordingly in a timely manner.

When kidney function first begins to improve, creatinine levels can continue to increase, but at a slower rate. As is illustrated on the left of Fig. 10.11, the traditional plot of creatinine levels versus time does not clearly convey the early improvements in kidney function (at point 4) that can signal the need to change treatment strategy.

This change for the better becomes clear by displaying the data not as a time series, but rather as a representation of the trajectory of creatinine levels versus time – a phase plot (Δ[Creatinine]/Δt versus [Creatinine]) as shown on the right of Fig. 10.11. Here it is quite clear that improvement begins at point 4. Most clinicians using the traditional plot would recognize improvement of function at point 9 thereby delaying the needed adjustments in the patient's treatment and possibly doing harm.[21]

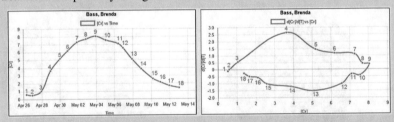

Fig. 10.11 The FHIR app discussed in this demonstration app created this pair of plots. The numbers along the plots correspond in time. In both plots, creatinine increases over time through point 9 and then begins to decrease. A traditional plot of creatinine versus time is on the left. On the right is a novel phase plot of creatinine level (X-axis) versus the first derivative of creatinine level divided by the first derivative of time (Y-axis). The phase plot makes it far more obvious that the rate of creatinine increase is decreasing after point 4. This signals the onset of renal function improvement much more clearly than can be appreciated in the traditional plot. (Courtesy Dr. Tim Buchman)

(continued)

---

[21] https://www.ncbi.nlm.nih.gov/pubmed/17622898

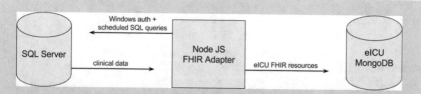

Fig. 10.12 The eICU FHIR adapter architecture. Clinical data (on the left) is pulled at regular intervals, formatted into FHIR resources, and placed into an eICU specific data mart (MongoDB on the right). (Courtesy Dr. Tim Buchman)

Early in this book we discussed the clinical oversight provided to ICUs by an eICU at Emory. These graphs were produced by a FHIR app developed by a team of Georgia Tech students as a class project working under the direction of Cheryl Hiddleson RN, Director, Emory eICU Center and Timothy G. Buchman, Ph.D., M.D., Director, Emory Critical Care Center and Professor of Surgery and Anesthesiology, Emory University School of Medicine.

Understanding how this was done provides a simple illustration of what it means to transform clinical data in a traditional database into FHIR resources. In this case the data was stored in a Philips eICU system[22] used by highly trained nurses and physicians to monitor care in the ICU and provide important oversight to make sure nothing gets overlooked. An example might be during a period of time when the ICU staff is focused on the care of a particularly critical patient.

As shown in Fig. 10.12, the student's FHIR adapter software did scheduled queries of the SQL database in the eICU system looking for new creatinine entries.

A FHIR REST API queries the data mart and responds with the JSON FHIR resources that the FHIR app expects. This project also illustrates that the clinical systems that can utilize FHIR apps may not be EHRs but could be virtually any system that stores a repository of clinical data.

Demonstration FHIR App: Sepsis Prediction in the ICU

The practice of medicine is largely about recognizing and appropriately responding to changes in patient status. We just saw an example of how a FHIR app could help physicians recognize a change in kidney function sooner than might otherwise be possible. However, no matter how timely that recognition is, it is still a response to something that has already happened. Might it be possible to predict the future with respect to changes in a patient's status?

Earlier we discussed sepsis as a severe, life threatening and common condition particularly among patients in intensive care units (ICUs). The recognition and treatment of sepsis sooner has been shown to reduce mortality. A 2006 study of over 2,700 patients with septic shock (the most severe form)

(continued)

[22] https://www.usa.philips.com/healthcare/product/HCNOCTN503/eicu-program-telehealth-for-the-intensive-care-unit

in 14 intensive care units in a mix of academic and community hospitals in Canada and the United States concluded that "Each hour of delay in antimicrobial administration over the ensuing 6 hours was associated with an average decrease in survival of 7.6%."[23] However, the clinical recognition of sepsis is difficult because the symptoms "may range from non-specific or non-localised symptoms (e.g., feeling unwell with a normal temperature), to severe signs with evidence of multi-organ dysfunction and septic shock."[24]

Predictive analytics uses a variety of statistical techniques from predictive modelling, machine learning, and data mining to analyze known data to make predictions about future or otherwise unknown events. Given the value of earlier recognition of sepsis a great deal of attention is being given to using predictive analytics to help make the diagnosis earlier. Since the analytics depends on current and past clinical data on the patient, an EHR integrated FHIR app would be a very attractive platform for a predictive sepsis tool.

Previously, we briefly discussed Sepsis Watch, a FHIR app that predicts changes in six lab values that can be indicators of the onset of sepsis and alerts physicians to determine the current values. The Sepsis Monitor app was developed by a group of Georgia Tech students mentored by Dr. Shamim Nemati, Assistant Professor in the Department of Biomedical Informatics at Emory School of Medicine. The goal was a cross-platform sepsis prediction and alert application that would work with any FHIR enabled EHR. Given the critical importance of timeliness the FHIR app would interface with the EHR in real-time to monitor multiple patients, calculate a sepsis score based using a machine learning algorithm and alert the care team when the scores suggest that sepsis is developing using a visual interactive interface. The app would also record data for analysis to improve the algorithm over time.

As shown in Fig. 10.13, inputs into the algorithm include heart rate, body temperature, blood oxygen saturation and blood pressure and over 30 other variables but not all of these inputs are required for the tool to function. Missing data is estimated using population means from ICU patients in MIMIC III. MIMIC III is an openly available dataset developed by the MIT Lab for Computational Physiology, comprising de-identified health data associated with ~40,000 critical care patients. It includes demographics, vital signs, laboratory tests, medications, and more.[25]

As we discussed in the previous demonstration app, an ICU can be a busy place particularly if one of its patients is in distress. There is always the risk that something will be overlooked so a dashboard that clearly flags any patients who are at high risk for sepsis would be of potential value. The app provides just this, as shown in Fig. 10.14, and each patient is given a sepsis score of between 0 (low risk) and 1 (high risk of sepsis). Patients are listed in descend-

(continued)

---

[23] https://www.ncbi.nlm.nih.gov/pubmed/16625125

[24] http://bestpractice.bmj.com/topics/en-gb/245/diagnosis-approach

[25] https://mimic.physionet.org/

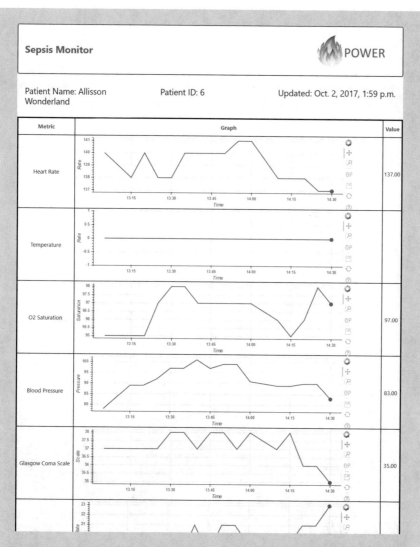

Fig. 10.13 The Sepsis Monitor FHIR app retrieves around 35 clinical variables from the simulated EHR. This is done using the MIT MIMIC ICU dataset that was mapped to OMOP and is therefore accessible as FHIR resources on the Georgia Tech HDAP server. Each variable is graphed and the most recent value is presented in the right most column. As shown in Fig. 10.14, Allisson Wonderland is at greatest risk of sepsis of the patients being monitored. (Courtesy Dr. Shamim Nemati)

ing order of sepsis risk further minimizing the likelihood of anything being overlooked. Allisson Wonderland, the patient whose physiologic variables are shown in Fig. 10.13, had the highest risk score at the time this display was captured but that can change over time so continuous monitoring is critical in these patients.

(continued)

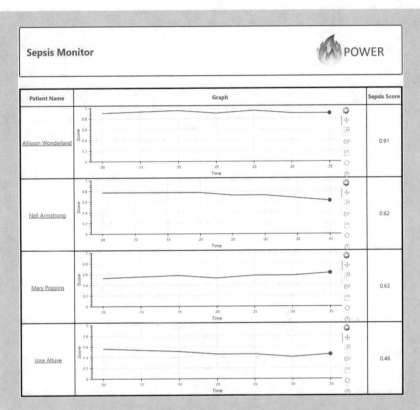

Fig. 10.14 Using the clinical variables shown in Fig. 10.13, the Sepsis Monitor FHIR app calculates a Sepsis Risk Score on a scale of between 0 (low risk) and 1 (high risk of sepsis). Allisson Wonderland is at greatest risk of sepsis of the patients being monitored at the time this display was captured but these are very sick patients who can be very unstable so this can change rapidly illustrating the value of a continuously updated dashboard. (Courtesy Dr. Shamim Nemati)

The details of the calculation of the Sepsis Risk Score are very technical. In this case it is done using a modified Weibull-Cox proportional hazards model that is one of the most popular forms of parametric regression models because it provides estimates of the hazard function, as well as coefficients for the variables that are possibly predictive of the outcome under study. A hazard function can be thought of as a measure of risk: the greater the hazard over some specified period of time, the greater the risk of an event in this time. In this case the time period is 4 h and the hazard function represents the risk of sepsis. The model has a prediction accuracy of over 80% which is felt to be sufficient for practical use in actual ICU patient care.[26]

---

[26] Joel R. Henry, Dennis Lynch, Jeff Mals, Supreeth P. Shashikumar, Ashish Sharma, and Shamim Nemati, A FHIR-Enabled Streaming Sepsis Prediction System for ICUs, 2018 40th Annual International Conference of the IEEE Engineering in Medicine and Biology Society (EMBC), Honolulu, Hawaii, July 17–21, 2018. – under review.

10.13 Recap

This chapter concludes Part III of the book. In the final part we will conclude by discussing some important and more dynamic and rapidly changing applications of health informatics.

Earlier in the book we discussed the need to enhance and extend functionality of existing health IT systems. SMART on FHIR provides a scalable, open framework that uses widely adopted solutions for identity authentication and authorization to access specific data. It accesses that data using the FHIR standard that is becoming a ubiquitous feature of the major electronic health record systems. This has led to the long sought goal of app stores for healthcare that support apps that can be relatively easily tuned to work with those EHRs. This facilitates innovation both because it substantially reduces the development effort to create cross platform tools and because it provides even startup companies with more efficient access to the potential end users.

We have seen a number of examples of the collaborations between end users and app developers that are becoming far more common as a result of this new ecosystem. We have also seen examples of the kind of innovative work that new entrants into the commercial FHIR app ecosystem are creating. Of course, innovations no matter how clever they may be, must be accepted by end users. The tight integration that SMART on FHIR facilitates between apps and the underlying EHR goes a long way toward solving the workflow and process issues that have long retarded provider acceptance of innovative solutions such as the early clinical decision support tools we have discussed.

It is probably too early to say that FHIR and SMART on FHIR have succeeded in transforming informatics into a practical everyday tool for providers and their patients but there is now ample reason to be optimistic that this may well be the case.

Part IV
New Frontiers

Chapter 11
mHealth

11.1 The Role of Patients in Preventing and Managing Chronic Disease

Many of us can think back to a key role a teacher played in shaping our world-view or our career. In my case, one such person was Dr. Hiram Curry, a general practitioner who retrained in neurology and then founded the first academic Family Medicine Department within a US based medical school. Its clinic has many of the characteristics we now associate with a patient centered medical home. He introduced me to Dr. Larry Weed's seminal book *Medical Records, Medical Education and Patient Care* and through it I became an advocate of his idea of a Problem Oriented Medical Record. I joined Dr. Curry's effort and led the development of one of the first ambulatory EHR systems. Dr. Curry taught me many things but one in particular is relevant to this topic. He used to caution his residents that they would see patients for a few minutes every so often and think that what *they did* in those few brief encounters would make all the difference. He was fond of saying that *what the patients did* between those encounters would actually make that difference.

We know that chronic disease accounts for most healthcare costs. We also know that a root cause of chronic diseases is often behavioral with poor diet, smoking and lack of sufficient physical activity being key health determinants. We also know that, once patients get chronic disease, their ability to positively change these negative behaviors and to comply with the prescribed treatments are the key drivers of a good outcome.

To understand how their patients were doing, conscientious physicians often asked them to maintain logs of diet, activities or treatment compliance. Today a new domain of health informatics, often called mHealth, seeks to use data from mobile apps, standalone or wearable devices and other sensors to understand health and health states, to inform health care actions and to change behavior in order to prevent disease onset. It also uses the data to promote behavior change once disease develops (e.g., medication adherence) and, increasingly, it is being used to provide clinical treatment (e.g., virtual physician visits, disease management or cognitive behavioral therapy).

© Springer International Publishing AG, part of Springer Nature 2018
M. L. Braunstein, *Health Informatics on FHIR: How HL7's New API is Transforming Healthcare*, https://doi.org/10.1007/978-3-319-93414-3_11

11.2   Does mHealth Produce Positive Results?

The World Health Organization (WHO) says that the "use of mobile and wireless technologies to support the achievement of health objectives (mHealth) has the potential to transform the face of health service delivery across the globe. A powerful combination of factors is driving this change. These include rapid advances in mobile technologies and applications, a rise in new opportunities for the integration of mobile health into existing eHealth services, and the continued growth in coverage of mobile cellular networks. According to the International Telecommunication Union (ITU), there are now over Five billion wireless subscribers; over 70% of them reside in low- and middle-income countries. The GSM Association reports commercial wireless signals cover over 85% of the world's population, extending far beyond the reach of the electrical grid."[1]

Does mHealth actually improve outcomes or lower costs? With all the excitement about sophisticated tools and apps the results to date may be surprising to many. A 2015 review article in the *Journal of Medical Internet Research* focused on mHealth as a tool to increase adherence in chronic disease management. The authors found 107 papers that met their criteria. Of these, 41 were randomized clinical trials that measured the effects of mHealth on disease-specific clinical outcomes in conditions such as diabetes or chronic cardiovascular and lung diseases. Sixteen of these studies (39%) reported significant improvements. Based on this, the authors say "There is potential for mHealth tools to better facilitate adherence to chronic disease management, but the evidence supporting its current effectiveness is mixed."[2]

A review of 75 clinical trials published in early 2013 was more positive. It concluded that "Text messaging interventions increased adherence to ART [antiretroviral adherence] and smoking cessation and should be considered for inclusion in services. Although there is suggestive evidence of benefit in some other areas, high quality adequately powered trials of optimised interventions are required to evaluate effects on objective outcomes."[3]

That text messaging (SMS) was the most proven intervention in the first study is interesting. In the more recent study it represented 40% of the mHealth activity was the most commonly used tool and the primary platform. Text messaging can facilitate patient-provider communication, medication reminders, and data collection and exchange on disease-specific measurements, as well as deliver patient education and motivation. It has the advantage of working on virtually all cell phones. This can be particularly important in developing countries where lower cost phones may be more common.

The next most common technology was smartphone apps for applications such as helping diabetes patients remember to check symptoms, maintain a food diary, or connect to educators in real time.

---

[1] http://www.who.int/goe/publications/goe_mhealth_web.pdf
[2] https://www.ncbi.nlm.nih.gov/pmc/articles/PMC4376208/
[3] http://journals.plos.org/plosmedicine/article?id=10.1371/journal.pmed.1001362

Next were standalone wireless or Bluetooth-compatible devices or those such as a blood glucose meter connected to a phone for automatic transfer of data for review by a health care provider. In some systems, measurements that fall outside of the target range trigger alerts.

11.3 mHealth Data Quality

Data quality is yet another mHealth challenge. A study in the December 2016 issue of the respected journal *Health Affairs* evaluated 137 patient-facing mHealth apps that were highly rated by consumers and recommended by experts and that targeted high-need, high-cost populations based on a long list of medical conditions. The authors found that "consumers' ratings were poor indications of apps' clinical utility or usability and that most apps did not respond appropriately when a user entered potentially dangerous health information." They further found that "very few apps focused on providing guidance based on user entered information or support through social networks, or on rewarding behavior change – functionalities likely to be useful to relatively more engaged patients." The authors also note that "data privacy and security will continue to be major concerns in the dissemination of mHealth apps". This is an interesting but, unfortunately, not yet openly available study.[4]

Another aspect of mHealth data quality is proper use of in-home devices by patients. They may incorrectly wear blood pressure cuffs. Patients may hold a book or some other object, steady themselves against a wall or other object or wear substantially different clothing as they weigh themselves. These issues clearly point to the need for careful training and devices that, to the extent possible, can detect such situations and coax patients to use proper technique.

Finally, even if devices are properly used, they may produce varying results. Keep in mind that much mHealth data is calculated and not directly measured. Step counts are a good example of this. Phones and wearable devices have accelerometers that measure changes in the user's movements. This stream of data must be interpreted to report on step counts and that is subject to error. A Swedish study evaluated the accuracy of six free pedometer applications for three different cell phones that were worn at three different positions. 10 subjects walked 200 steps with each application, cell phone, and cell phone position. The study concluded that "Only one application and cell phone combination showed a good accuracy with reasonable low standard deviation, especially in one of the cell phone positions. The majority of applications evaluated in this study, did not show high accuracy."[5]

---

[4] https://www.healthaffairs.org/doi/abs/10.1377/hlthaff.2016.0578?url_ver=Z39.88-2003&rfr_id=ori%3Arid%3Acrossref.org&rfr_dat=cr_pub%3Dpubmed

[5] https://ac.els-cdn.com/S221201731200518X/1-s2.0-S221201731200518X-main.pdf?_tid=62c3acb6-11a3-11e8-9cba-00000aab0f6c&acdnat=1518625587_91f9c8f2710f1d841c2fae73c5ff57cd

While the accuracy of the technology may be improving it can depend to a great deal where on the body measurements are taken. A more recent 2015 study published in the *Journal of the American Medical Association* asked healthy adults aged 18 years or older to walk on a treadmill set at 3.0 mph for 500 and 1500 steps. The subjects wore 10 top selling apps and devices on various positions on the body. The study concluded that "the relative difference in mean step count ranged from −0.3% to 1.0% for the pedometer and accelerometers, −22.7% to −1.5% for the wearable devices, and −6.7% to 6.2% for smartphone applications."[6]

11.4   The FDA

Here in the US, the Food and Drug Administration (FDA) is responsible for regulating medical devices. The details are complex but the level of FDA oversight generally increases with the perceived patient risk presented by the device. Devices that perform the same function as already approved 'predicate' devices and do not introduce new risks are generally subject to minimal regulation (via the 510 K approval process). Novel devices may need to go through a far more complex, expensive and time-consuming process.

What about medical apps? Given finite resources and the scale of app development it was clear that the FDA had to draw a line to determine which apps it would even consider. In the summer of 2017 the agency announced its Digital Health Innovation Plan to modernize and streamline its Center for Devices and Radiological Health's (CDRH) approach to regulating digital health devices.[7]

The plan is intended to:

- Issue guidance to provide clarity on the medical software provisions of the twenty-first century Cures legislation;
- Launch a new pilot precertification program to work with industry to develop a new approach to digital health technology oversight (FDA Pre-Cert for Software); and
- Increase the FDA's digital health unit's capability and capacity.

A key element of the plan that has worked well in the author's view at CDC and CMS is an Entrepreneurs in Residence program to bring thought leaders and others with experience in software development into the CDRH.

---

[6] https://jamanetwork.com/journals/jama/fullarticle/2108876

[7] https://www.fda.gov/downloads/MedicalDevices/DigitalHealth/UCM568735.pdf

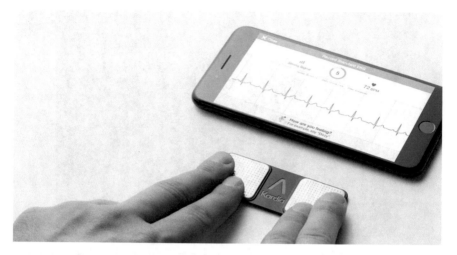

Fig. 11.1 AliveCor's Kardia Mobile device detects the electrocardiogram (ECG) heart trace and sends it to the company's smartphone app using inaudible ultrasonic signals after requesting access to the microphone. (Courtesy AliveCor)

11.5   AliveCor

We will now consider an example of a mHealth device that the FDA does regulate. AliveCor's Kardia Mobile device (Fig. 11.1) detects the electrocardiogram (ECG) heart trace and sends it to the company's smartphone app using inaudible ultrasonic signals after requesting access to the microphone. The company says this communications approach reduces battery consumption by around 92% versus Bluetooth and provides "much higher resolution data". However, it does require the device to be no more than a foot away from the phone.

Atrial fibrillation (AF) is the most common cardiac arrhythmia, affecting an estimated 33.5 million individuals worldwide and is an important risk factor for stroke. It may be associated with a third of strokes and is often not diagnosed before a stroke occurs. Better screening for AF in patients at greater risk (the risk increases with age and a number of other known factors) could lead to treatment before a stroke occurs. A number of journal articles (including one in the prestigious journal *Circulation*) report on success in using the AliveCor device and other competing devices to screen for atrial fibrillation.[8] This is an interesting an early indication of how important mHealth will likely become in the future practice of medicine.

AliveCor has released KardiaBand™ (Fig. 11.2) a special band for the Apple Watch. The company's SmartRhythm software acquires data from the watch's heart rate sensor and accelerometer and uses a neural network (a form of machine learning that can progressively improve performance on tasks by considering examples, generally without task-specific programming) to compare actual heart rate changes to what it expects based on the user's activity.

---

[8] http://circ.ahajournals.org/content/early/2017/08/28/CIRCULATIONAHA.117.030583

Fig. 11.2 AliveCor's KardiaBand provides similar functionality to the Kardia device using a special Apple Watch band and can take advantage of activity data measured by the phone to correlate it with heart rate helping it to recognize rates that may signal atrial fibrillation. This is an interesting early example of using more than one mHealth metric to do potentially useful analysis. (Courtesy AliveCor)

Heart rate is important in screening for atrial fibrillation. The heart has four chambers – smaller atria above and larger ventricles below. In a normal heart the beat originates in the right atrium and the rate is well controlled and varies with activity. Atrial fibrillation is an irregular and often rapid heart rhythm that can cause the heart rate to increase to well above normal.

When the network sees a pattern of heart rate and activity that it does not expect, it notifies the user to take an ECG using the band. The user can see the ECG on the watch and can forward it to their physician. At present the company says that SmartRhythm can distinguish a normal heart rhythm from atrial fibrillation and from other abnormal rhythms. It appears that the company hopes to increase the number of specific arrhythmias the system can detect and classify accurately.[9] There is an interesting brief video from AliveCor's founder that suggests the company's future direction and the growing importance of analytics in healthcare.[10]

11.6   Device and App Interoperability

Unsurprisingly, interoperability of medical devices is a challenge both within and outside of hospitals. A 2012 Association for the Advancement of Medical Instrumentation (AAMI) survey of 1900 U.S. hospitals found that interoperability issues were first and second on the list of medical device challenges. Specifically, they were integrating medical devices and systems into the hospital's network (cited

---

[9] https://www.alivecor.com/technology/

[10] https://vimeo.com/255102203

by 72% of respondents) and integrating device data into EHRs (cited by 65% of respondents).[11]

When we looked at HealthVault you saw a plethora of devices used by patients that could upload data to this PHR. Each of them typically has its own proprietary data format so HealthVault provided value by bringing these formats so that data was brought together in one record. It provided additional value by making that mHealth data available along with other data collected from EHRs or documented by the patient. To give but one example of this added value, a patient could potentially correlate increased activity with better control of their hypertension or diabetes. An app that takes advantage of the availability of these multiple data sources would be the ideal way for patients to do that.

In the past, individual device data was often only available via a portal maintained by its manufacturer and there might have been a charge to use it. The Continua Health Alliance, founded in 2006, is an international non-profit, open industry group of nearly 240 healthcare organizations and vendors. Its goal is to establish mHealth interoperability in three major categories: chronic disease management, aging independently, and health and physical fitness.

Case Study: The Personal Connected Health Alliance

The Personal Connected Health Alliance (PCHAlliance) is a non-profit organization formed by the huge Healthcare Information and Management Systems Society (HIMSS), Continua, and the mHealth Summit to "realize the full potential of personal connected health". It now publishes and promotes the global adoption of Continua's open mHealth interoperability framework.

In late 2017 PCHAlliance released a new version of the Continua Design Guidelines (CDG) that, for the first time, uses the FHIR specifications. According to PCHAlliance "this includes support for personal health data from 26 vital signs sensors and 40 health, medical and fitness capabilities enabling hundreds of different product types that can now be certified to the CDG today – and all with a direct path to the EHR. This includes many products for telehealth and telemonitoring of chronic diseases, including diabetes, heart failure, hypertension and COPD, as well as health and fitness measures."[12, 13]

The Continua Personal Health Device Data Implementation Guide v0.1.0 is posted on the HL7 site and it describes personal health devices (PHDs) as those that are "mostly used in home-care contexts and include Continua-certified devices such as glucose meters, blood pressure cuffs, weight scales,

---

(continued)

---

[11] http://www.aami.org/productspublications/pressreleasedetail.aspx?ItemNumber=4177

[12] http://www.pchalliance.org/news/new-continua-design-guidelines-enable-integration-patient-generated-data-electronic-health

[13] http://www.incisor.tv/fb/01122017/mobile/index.html#p=9

thermometer, etc. The PHD 'information' in this context means both the measurements taken by the PHD and data about the PHD itself. The PHD data includes characteristics, operational status and capabilities for the device, such as the serial number, manufacturer name, firmware revision, etc."[14]

It also describes a personal health gateway (PHG) that is responsible "for receiving and decoding the information from the PHD and either generating and uploading the FHIR resources or providing sufficient information to a backend server such that the backend server can generate the FHIR resources."

Patient information, such as demographic and administrative information about the patient is usually not provided by PHDs and must be supplied and associated with PHD data. In a FHIR scenario it would be represented as a FHIR Patient resource. Since PHD data may be communicated over a public network PHGs may be supplied with an opaque and unique 'key' that only the health care provider can link to a patient avoiding the need to store or communicate data that would be considered personally identifiable and therefore HIPAA protected.

As would be expected the PHD data is instantiated in FHIR Observation resources and is coded using LOINC. Note that Continua is now an IEEE recognized standard (11073) so there is also a reference to that in the FHIR resource. Figure 11.3 is taken from the implementation guide and provides a simple example of this representing a pulse rate measurement from a patient operated blood pressure cuff.

Understanding the context of information recorded by devices operated by the patient in their home is critical. One aspect of that is information about the device itself. As shown in Fig. 11.4, there is a FHIR compliant JSON DeviceComponent resource specification for this purpose. This is a blood pressure cuff device that might well have obtained the heart rate measurement illustrated in Fig. 11.3. Note that the manufacturer, the device model and serial numbers and the firmware version can be reported but all these elements are optional.

Continua CODE for Healthcare: PCHAlliance's longer term goal is to make it easier for product developers to incorporate the Continua ecosystem into their commercial offerings by using a Health and Fitness Server (HFS) though a Free Open Source Software project called CODE for Healthcare. CODE stands for Continua Open Development Environment. The Health and Fitness Server will securely receive observations from Continua devices without the need to create software that performs the reception and translation functions documented for the services interface in the Continua Guidelines.

(continued)

[14] http://hl7.org/fhir/uv/phd/2018Jan/index.html

The proposed system will also be designed, implemented and tested such that it addresses basic regulatory requirements in terms of software development processes, a requirement for FDA 510 K device certification.

The first phase of this effort will create a baseline code framework for a Health and Fitness Server that incorporates a FHIR receiver component, an OAuth security framework, an applications integration bus, an underlying database, and a web service exporting a REST API for administration of the system. The majority of the work of this first phase is focused on establishing the extensible framework that will support future open source development.

With the initial framework in place, additional functional components will be implemented as well-defined, independently deployable microservices (often an HTTP API). Currently the Health and Fitness Server has two services associated with FHIR. The first ensures that the entity seeking to perform the FHIR operation has sufficient access rights. The second is the FHIR server itself.

Future work may include adding Internet of Things (IoT) technologies, existing Continua certified device classes, or other yet to be defined capabilities. This approach is intended to be both cost effective for PCHA and to encourage outside developers to engage in the project.

The major goals for the project are: (1) That an IT professional can download the software and quickly (within a day) have a proof of concept system up and running, and (2) that a software development engineer could identify and integrate the specific Continua functionality needed from the CODE for Healthcare into a commercial product and have the test, license, and intellectual property artifacts needed to be comfortable with using the CODE for Healthcare component in the product.

(continued)

```
{
    "resourceType":"Observation",
    "meta":{
        "profile": [
            "http://pchalliance.org/phdfhir/StructureDefinition/PhdNumericObservation"
        ]
    },
    "identifier": [
        {
            "value":"sisansarahId-urn:oid:1.2.3.4.5.6.7.8.10-01040302f0000000-149546-20171212091343-48-264864"
        }
    ],
    "status":"final",
    "category": [
        {
            "coding": [
                {
                    "system":"http://hl7.org/fhir/observation-category",
                    "code":"vital-signs",
                    "display":"Vital Signs"
                }
            ]
        }
    ],
    "code":{
        "coding": [
            {
                "system":"urn:iso:std:iso:11073:10101",
                "code":"149546",
                "display":"MDC_PULS_RATE_NON_INV"
            },
            {
                "system":"http://loinc.org",
                "code":"8867-4",
                "display":"Blood Pressure cuff heart rate"
            }
        ]
    },
```

Fig. 11.3 The JSON representation of a FHIR Observation resource as it might be output by a Continua compliant personal health gateway (PHG). The LOINC code of 8867-4 is 'heart rate' but there are a large number of LOINC codes around this particular measurement. You might want to visit https://search.loinc.org/ to see how complicated LOINC coding can be and figure out what code you feel would be optimal for this resource given the source of the value. Note as well that Continua is now an IEEE recognized standard (11073) so there is a reference to that as well. (Courtesy Continua)

(continued)

```
"productionSpecification": [{                    // All display elements are optional but encouraged
        "specType": {
                "coding": [{
                        "system": "urn:iso:std:iso:11073:10101",
                        "code": "531970",
                        "display": "MDC_ID_MODEL_MANUFACTURER: Manufacturer name"
                }]
        },
        "productionSpec": "Renesas Electronics"
},
{

        "specType": {
                "coding": [{
                        "system": "urn:iso:std:iso:11073:10101",
                        "code": "531969",
                        "display": "MDC_ID_MODEL_NUMBER: Model number"
                }]
        },
        "productionSpec": "Synergy-12345-Demo"
},
{

        "specType": {
                "coding": [{                                    // MDC coding system first
                        "system": "urn:iso:std:iso:11073:10101",
                        "code": "531972",
                        "display": "MDC_ID_PROD_SPEC_SERIAL: Serial number"
                },
                {
                        "system": "http://hl7.org/fhir/specification-type",
                        "code": "serial-number",
                        "display": "Serial number"
                }]
        },
        "productionSpec": "13456-BPM-BTLE"
},
{

        "specType": {
                "coding": [{
                        "system": "urn:iso:std:iso:11073:10101",
                        "code": "531976",
                        "display": "MDC_ID_PROD_SPEC_FW: Firmware revision"
                },
                {
                        "system": "http://hl7.org/fhir/specification-type",
                        "code": "firmware-revision",
                        "display": "Firmware revision"
                }]
        },
        "productionSpec": "1.0.0"
```

Fig. 11.4 The JSON representation of a FHIR DeviceComponent resource as it might be output by a Continua compliant personal health gateway (PHG). This is a blood pressure cuff that might have obtained the pulse rate reported in the Observation resource shown in Fig. 11.3. The manufacturer, the device model and serial numbers and the firmware version can be reported, but all are optional. (Courtesy Continua)

(continued)

11.7   Commercial mHealth Data Integrators

Validic and Human API are two (but certainly not the only two) interesting companies that have each developed an interoperability platform for mHealth data from a
wide variety of sources aimed at both patient care and wellness. Both offer REST
API's to access their normalized data in JSON formats. As we will discuss, Human
API has now greatly expanded their data store beyond mHealth in an effort to offer
a wider view of health.

Case Study: Validic
The company says its goal is helping its clients turn disparate mHealth data
into actionable insights. Toward this end, its Inform solution seeks to provide
more facile access to health data from nearly 400 clinical and consumer data
sources with pre-authorization by the patient or consumer whose data is being
shared. Inform's major components include the data connectivity platform,
mobile libraries, and rules engine.

The connectivity platform integrates data in three ways: via direct API
connections built by Validic to device vendors, such as with Fitbit and Garmin;
Validic Connect, which enables device vendors to build directly to the Validic
API; and Mobile Bluetooth, software development kit (SDK), and optical
character recognition (OCR) Connections via Validic's mobile libraries. Once
the data is extracted, the platform converts it into a standard JSON format.
The company says the platform is HIPAA compliant. All data stored in and
passed through the platform is de-identified according to the CMS Safe
Harbor rule.

Validic's mobile libraries enable healthcare organizations to integrate data
from Bluetooth Smart devices, Apple's HealthKit, and data-locked legacy
devices (via VitalSnap®, described next) into their existing iOS and Android
mobile applications. Mobile Libraries provide embeddable code for both legacy and Bluetooth clinical devices. Data acquired from Validic Mobile is
available to clients via the data connectivity platform.

VitalSnap allows users to point the phone's camera at non-connected
devices like blood glucose meters, blood pressure monitors, or pulse oxime-
ters and capture the data from their screens without actually taking a picture.
This data can then be sent to healthcare providers and other Validic customers
the same way data from a connected device would be.

The rules engine helps care managers and program administrators track
population and individual engagement, as well as program adherence.
Adherence rules can trigger notifications or flags when a member or patient
has not submitted data over a specified amount of time. To help manage what
could be very large, continuous streams of real-time data, rules can help set
trigger-based actions and alerts. The rules are set and triggered via the data
connectivity platform.

(continued)

The data connectivity platform offers three ways to integrate data: via a REST API, a Streaming API, or a combination of the two. As shown in Fig. 11.5, the Inform REST API supports Validic customers that are more interested in summary data and less interested in real time access. This allows a healthcare organization to pull data once it is made available in the platform. The API is designed to provide data for a specific user for a specific time period. The Inform Streaming API gives developers access to all of their data through a continuous stream in near-real time without any of the overhead associated with polling a REST endpoint. The target is customers who want to consume and process all of their users' data as it is made available from the device manufacturer.

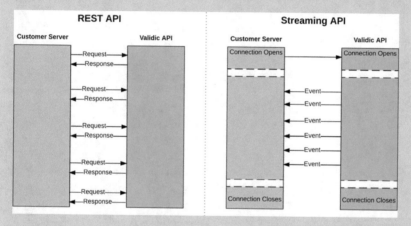

Fig. 11.5 The Inform REST API supports Validic customers that are more interested in pulling data for a specific user for a specific time period once it is made available in the platform. The Inform Streaming API provides access to all data through a continuous stream in near-real time without any of the overhead associated with polling a REST endpoint. (Courtesy Validic)

Case Study: Human API

While Human API began with a focus on mHealth data integration it now says that its mission is to put consumers in control of all of their health data and make it easier for them to share it with whatever entities they feel can provide them with better care and services. The company also seeks to help those enterprises access, understand, and leverage their consumers' rich health information. To accomplish both goals the company has created what it describes as a large, distributed network of integrations and a simple, on-demand way for everyone to exchange and use the health data they need. Its current customer segments include digital health, clinical research, life insurance, healthcare providers, and health insurance.

The use cases powered by Human API include data-driven chronic condition management tools, streamlining the new patient intake process, accelerating the underwriting process for life insurance and expediting the clinical trial eligibility determination process.

Customers integrate Human API into their application(s) by adding the company's simple code segments to them. This might power a "Connect my data" button that could be presented to patients during an on-boarding/registration flow, as an in-app notification, within a settings section, or via email, among other possibilities.

The company announced at HIMSS 2018 that it was making the new CMS Blue Button 2.0 API available through the Human API data network. The CMS Blue Button 2.0 API uses FHIR to access 4 years of claims data for 53 million Medicare beneficiaries.

The example of using Connect shown in Fig. 11.6 takes advantage of this new capability and illustrates a Medicare patient being asked to share their claims derived health data with a hypothetical health system where they are receiving care. The Connect experience is configurable to the context in order to facilitate the user finding their provider and authorizing Human API to retrieve their health data on their behalf.

The customer's users (e.g. a hospital's patients) find their providers via the company's ProviderGraph and authorize their health data to be retrieved and shared as they choose. The multi-dimensional tool is supported by a map of the relationships between healthcare providers, doctors, hospitals, and also has metadata associated with each (names, URLs, locations/addresses, phone or fax #s, etc.) to make search and identification easier, guide the users through registration or recovering credentials, and tying providers to data. Patients can choose to electronically share their data directly (e.g. making select source data available via PDF, CCD, etc.) or, as illustrated in Fig. 11.6, sharing their data electronically with a Human API customer/partner (who can pull the data on-demand via the API) (Fig. 11.7).

(continued)

The data pipeline begins with the retrieval of the data from sources that include patient portals, partnerships with data sources, health information exchanges and even offline chart retrieval and also by using APIs, including FHIR.

Data is securely stored and then passed to the company's data parsing system that understands common formats, including JSON, CCDs, and HTML and uses machine learning to translate them into the Human API internal data structures. This includes normalizing and enriching the data with data standards (RxNorm, NDC, SNOMED-CT, ICD 9/10, LOINC and CPT) and standardizing dosages, measurements, etc. where possible. The company says it has built its own ontology "that's inspired by, but not limited to the current FHIR specifications" with the goal of making software development easier for common use cases and to answer common questions.

In addition to mHealth data from apps, medical and wearable devices, the data includes clinical, claims and genomic information. Once it is available, customers are notified and the data is accessible via the company's API. Currently the company says its data network includes over 35,000 EHRs, labs, pharmacies, patient portals, devices, and mobile apps; its data store contains over 50 billion data points obtained from its access to over 230 million US consumers; and it processes over 50M million API calls daily

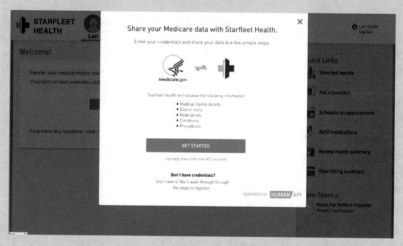

Fig. 11.6 An example of the integration of Human API's 'connect my data' button into their customers' system. Here, a Medicare patient is asked to share their claims derived health data via Human API with a hypothetical health system where they are receiving care. (Courtesy Human API)

(continued)

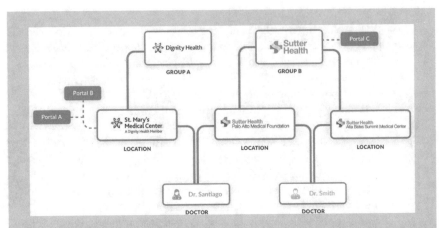

Fig. 11.7 The Human API ProviderGraph structure facilitates patients finding their data sources. (Courtesy Human API)

11.8   Open mHealth

Open mHealth is a non-profit organization that promotes greater access to mobile health data through open standard and tools. Like FHIR, the Open mHealth standard uses REST APIs to access data packaged according to JSON-based 'schemas' – essentially the equivalent of FHIR resources but with important differences because of the challenges of mHealth data. While the Open mHealth standard and the Continua standard for representing mHealth data have some similarities, they are currently distinct approaches.

Open mHealth also provides tools to:

- validate data
- pull in data from large and popular device manufacturers
- store data and share it securely with others
- process and visualize data

Figure 11.8 is an example of part of the Open mHealth Schema for blood glucose. As with FHIR it is formatted in JSON. As with Continua, this is a FHIR Observation resource. The important distinctions are the things included in Open mHealth but not in FHIR or in Continua. First, if you look at the start and end dates and times you see that the glucose "measurement" occurred over a 4-month period!

mHealth data has many unique challenges and this illustrates one of them – volume. A diabetes patient might document their blood glucose daily. They see their physician every 4 months. Does that physician want to see all those data points? Almost certainly not, so Open mHealth, as indicated in the last line of JSON in Fig. 11.5, might provide an average for the 4-month period. For blood pressure, this might alternatively be the maximum and minimum over some time-period.

```
{
    "blood_glucose": {
        "unit": "mg/dL",
        "value": 128
    },
    "effective_time_frame": {
        "time_interval": {
            "start_date_time": "2015-02-05T07:25:00Z",
            "end_date_time": "2015-06-05T07:25:00Z"
        }
    },
    "temporal_relationship_to_meal": "fasting",
    "temporal_relationship_to_sleep": "on waking",
    "descriptive_statistic": "average"
}
```

Fig. 11.8 This illustration of an Open mHealth Glucose schema illustrates that it offers calculated metrics to summarize what might be a large number of data points in a more clinically useful and efficient way. It also offers important context about the data collection that can substantially impact its clinical usefulness. (Courtesy Open mHealth)

Another mHealth challenge is data context – when and how was data collected? This should be clear for data obtained in a physician's office but it may not be when the patient is at home. Here the schema provides contextual information showing that the patient takes their glucose reading upon waking and before eating. This is usually what the physician tries to do when the patient visits their office. Absent this information it would be more difficult to interpret and use the average glucose value clinically.

11.9   Open mHealth Tools

In addition to the schemas, Open mHealth provides useful tools for developers in the mHealth space. One of these allows users to specify a dataset they want generated using YAML, a data serialization language designed to be both human readable and computationally powerful. A developer might, for example, specify that body weights are desired over a particular date range and with a specified starting and end value, standard deviation and minimum and maximum values. Open mHealth also provides a library of visualization tools to use once the data is generated.

Earlier we discussed Human API and Validic, two companies that offer proprietary platforms for integrating mHealth data from multiple sources. Open mHealth has developed Shimmer, an open source tool to pull health data from popular third-party APIs like Runkeeper and Fitbit and convert it into an Open mHealth compliant format. Just as FHIR does for EHR data, Shimmer 'isolates' developers from proprietary mHealth formats so they can write once to Open mHealth schemas and their apps should work with any mHealth data source supported by Shimmer. Currently the number of data sources is small compared to the commercial entities but Open mHealth says it is working to add more.

11.10   Apple's HealthKit

Earlier we discussed the importance of Apple's support of FHIR so that iPhone users could upload their medical record data to their phone. In that discussion we identified the ability to combine that data with mHealth data as potentially quite useful and important. In the AliveCor discussion we saw an example of that when its KardiaBand could supply ECG data to an app on the iPhone that could also access activity data to assess whether the heart rate was consistent with the patient's activity or, alternately, might be due to atrial fibrillation.

Apple's HealthKit, when combined with EHR data, offers the potential to take this idea to a new level. HealthKit stores data from iPhone and Apple watch apps and from some devices in an encrypted database called the HealthKit store. The data is only accessible via a Health app through which users can view, add, delete, and otherwise manage all of their health and fitness data. Importantly, as shown in Fig. 11.9, users can control fine-grained permissions for the sharing of each data

Fig. 11.9 Apple's HealthKit provides users with the ability to share data at a granular level and to control whether each data type can be read only or can be written to their record by a mHealth device. The granular nature of FHIR resources fits into a similar approach to user permissions. (Courtesy Apple)

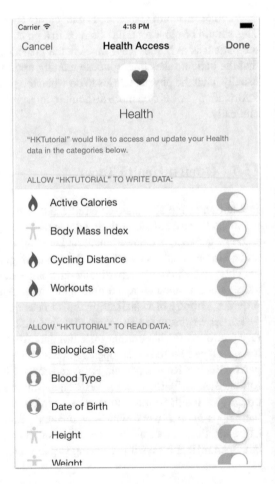

type. Users must explicitly grant each app permission to read and write data to the HealthKit store. Users can grant or deny permission separately for each type of data. For example, a user could allow an app to read their step count data but prevent it from reading their blood glucose level. To prevent possible information leaks, an app does not know about denials and, from the app's point of view, data for which it lacks read permission does not exist. The granular nature of FHIR resources fits well into this approach so one can easily envision how HealthKit would manage permissions for sharing of EHR data.

Once an app has been granted permission to access data in the store it can use Open mHealth's Granola tool to store it on a remote server for analysis or backup or to serialize the data that is only stored in the HealthKit store for a limited period of time. To accomplish this, Granola maps and validates the HealthKit API to Open mHealth JSON schemas.

11.11   Recap

If any chapter of this book is subject to almost immediate obsolescence it is this one! mHealth is truly the new frontier of healthcare. Hopefully, the chapter will provide you with the background to appreciate the many innovative ways that this technology has been applied and surely will be applied in the future. This is also an area where the interests of patients, providers and payers seem to be aligning and that will provide the long needed stimulus to more widespread application of mHealth technologies in order to help solve some of the fundamental challenges of healthcare delivery we introduced at the outset of this book.

Chapter 12
Public and Population Health

12.1 Introduction

In the clinical setting, the focus is traditionally on individual patients after they become sick or injured. In contrast, public and population health focus on groups of patients or the public at large. These can be as small as a local neighborhood, or as big as an entire country or region or the world. Instead of diagnosis and treatment of individuals, the goal may well be prevention of disease, identification of risk or protective factors, or promotion of the health of communities as a whole. We will begin this chapter with a discussion of public health and then shift to population health.

12.2 The Roots of Public Health

Since its outset public health has been about collecting and using data to understand health and disease. To a large degree the development of public health is intertwined with the roots of the International Classification of Diseases. They both began with trying to understand why people die and have now expanded to understanding disease and its causes and patterns at a population level.

Public health can be traced back to the mid-fifteenth century in northern Italy, a period of the great plague. For the first time, death certificates from the victims of the plague were used as a source of data for analysis. In the mid-1600s John Graunt, one of the first demographers, published *Observations on the Bills of Mortality*, an analysis of bills of mortality in London for the purpose of determining why children died. This was essentially the same idea: using mortality data to gain knowledge about health, at least at a population level, the precursor to today's public health.

The 1700s saw the beginning of the first national registration systems for births, deaths and certain diseases. England passed a national registration act in 1837.

© Springer International Publishing AG, part of Springer Nature 2018
M. L. Braunstein, *Health Informatics on FHIR: How HL7's New API is Transforming Healthcare*, https://doi.org/10.1007/978-3-319-93414-3_12

Massachusetts was the first state in the US to do the same in 1842. Not long thereafter, with this data becoming available, people saw the need to classify the information to achieve a degree of consistency and to organize it into a more manageable form. One long list would be unwieldy to work with so there needed to be some internal classification of the information. William Farr first proposed this in 1839 in the *First Annual Report* of the General Register Office (responsible for the civil registration of births, adoptions, marriages, civil partnerships and deaths in the UK and Wales) where he was working:

> The advantages of a uniform nomenclature, however imperfect, are so obvious, that it is surprising no attention has been paid to its enforcement in Bills of Mortality. Each disease has, in many instances, been denoted by three or four terms, and each term has been applied to as many different diseases: vague, inconvenient names have been employed, or complications have been registered instead of primary diseases. The nomenclature is of as much importance in this department of enquiry as weights and measures in the physical sciences, and should be settled without delay

Surprisingly, his well-reasoned idea received scant attention at the time. From today's perspective this is the foundation of the interoperability problem we're still dealing with almost 180 year later! Interestingly enough, the triggering event to making this happen was the Great Exposition of 1851 where various countries presented displays. This led to discussion of the need for standards for cross-country statistical comparison. This led to the idea of meeting internationally to create a uniform classification of diseases. The first meeting was in 1853, with a second in 1855 where the issue of how this data could be grouped was a topic.

Marc d'Espine proposed a grouping according to what could be termed the nature of diseases: gouty, herpetic, hematic, and so on. These are not terms we use to group diseases today (or, in most cases, use at all). William Farr proposed five groups according to the cause of diseases: epidemic diseases, constitutional (general) diseases, local diseases arranged according to anatomical site, developmental diseases, and diseases directly resulting from violence. In 1893, Alphonse Bertillon proposed the first reasonably clinically correct grouping. The categorical organization of the list of diseases we use today is not substantially different from his list.

Another story anticipates the modern use of geographically coded data in public health by over 150 years. At the time, most thought that cholera was spread by "miasma", or bad air. Dr. John Snow believed that drinking water was its cause although he did not know the mechanism since this was a few years before Louis Pasteur proposed the germ theory of disease causation.

In August 1854 a major outbreak of cholera occurred in the Soho section of London and 127 people on or near Broad Street died. By talking to local residents Snow identified the source of the outbreak as the public water pump on Broad Street (now Broadwick Street) at Cambridge Street. Despite having no scientific proof other than his 'epidemiologic' data about the patterns of illness and death among residents Snow persuaded the authorities to disable the well pump by removing its handle. By then residents had fled the area so the disease was already in decline making actual proof of Snow's theory problematic. However, Snow created a map to show how cholera cases were clustered around the pump. This is now known as a Voronoi diagram. He also did statistical analysis to illustrate the connection between the quality of the source of water and the number of cholera cases.

As a result of this and other work John Snow had a seminal influence on the future of public health.

12.3   Public Health Today

Since the early days of using mortality data to understand disease causality and progression, the mission and reach of public health has substantially expanded. In a landmark 1998 report *The Future of Public Health*, the Institute of Medicine defined public health as "what we as a society do collectively to assure the conditions in which people can be healthy."[1] Note the emphasis on collective action. This is a recognition that complex social factors have a powerful influence on our health and that we require disparate data sources to understand the impact of many determinants of health and disease. The World Health Organization, a globally recognized public health organization, defines health as "the state of complete physical, mental, and social well-being and not merely the absence of disease or infirmity."[2] Today, public health takes a broad view of the factors that influence health – where people live and work; the social norms of our culture; the types of healthcare we have access to; our exposure to diseases and stressors in our environment; among other influences. Thus, from an informatics perspective, public health agencies require disparate data sources to understand the impact of many determinants of health and disease.

The CDC has identified these ten essential roles for public health:

1. Monitor health status to identify and solve community health problems
2. Diagnose and investigate health problems and health hazards in the community
3. Inform, educate, and empower people about health issues
4. Mobilize community partnerships and action to identify and solve health problems
5. Develop policies and plans that support individual and community health efforts
6. Enforce laws and regulations that protect health and ensure safety
7. Link people to needed personal health services and assure the provision of health care when otherwise unavailable
8. Assure competent public and personal health care workforce
9. Evaluate effectiveness, accessibility, and quality of personal and population-based health services
10. Research for new insights and innovative solutions to health problems

---

[1] http://www.nationalacademies.org/hmd/Reports/1988/The-Future-of-Public-Health.aspx
[2] https://www.ncbi.nlm.nih.gov/books/NBK221233/

To accomplish these objectives, public health often partners with:

- Public health agencies at state and local levels
- Healthcare providers
- Public safety agencies
- Human service and charity organizations
- Education and youth development organizations
- Recreation and arts-related organizations
- Economic and philanthropic organizations
- Environmental agencies and organizations[3]

Interoperability is, of course, of great importance in public health, where the main mission is gathering accurate and timely data from providers and other sources usually across a large geographic area. FHIR could potentially facilitate this data aggregation and, as we will see in what follows, is increasingly seen as an important health informatics tool.

We will now look at two case studies that demonstrate how FHIR could be used in public health. The first focuses on the historic root question we just discussed, why people die. The second illustrates how the frontiers of public health are expanding in the new world that is rich in digital data and APIs to access that data.

12.4   Electronic Mortality Reporting

The CDC describes its mission as protecting America from health, safety and security threats wherever they originate. According to the CDC's publication *Physician Handbook on Medical Certification of Death*[4] the death certificate is the permanent record of the fact of death. In the United States there are 57 jurisdictions that are responsible for mortality registration so various state laws specify the required time for completing and filing the death certificate.

Again, according to the CDC, the death certificate provides important personal information about the decedent and about the circumstances and cause of death. It is the source for state and national mortality statistics that show the substantial changes in life expectancy and the causes of mortality over time, as shown in Fig. 12.1. It is used to determine which medical conditions receive research and development funding, to set public health goals, and to measure health status at local, state, national, and international levels.

There are a number of data sources but physicians are critical to accurately document the key data describing the cause of death. According to the Stanford University

---

[3] https://www.cdc.gov/stltpublichealth/publichealthservices/essentialhealthservices.html
[4] https://www.cdc.gov/nchs/data/misc/hb_cod.pdf

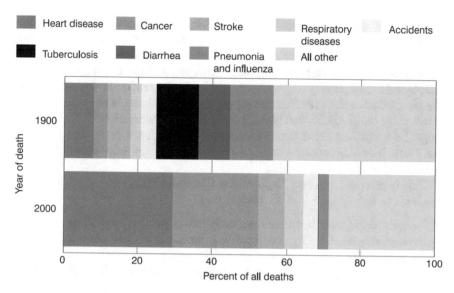

Fig. 12.1 Over the twentieth century there was a dramatic shift in the reasons for mortality from infectious diseases to the chronic diseases that now drive most healthcare. The analysis of death certificate data allows the CDC to inform us of trends such as this and to better align public policy with the changing needs of society. (CDC; https://blogs.cdc.gov/nchs-data-visualization/deaths-in-the-us/)

School of Medicine "60% of Americans die in acute care hospitals, 20% in nursing homes and only 20% at home."[5]

In a hospital setting it is not uncommon for the physician who pronounces the death of a patient and is expected to complete the Medical Certifier section of the death certificate to be unfamiliar with the patient. For the purposes of collecting important information on the causes of death this section of the form asks for the immediate cause of death as well as a sequential list of the conditions that led to the final event that caused death. For example, a patient may have died because of a pulmonary embolism (a clot that lodges in the lungs) but the causal sequence that led to this may have begun with chronic heart disease that led to a heart attack that caused congestive heart failure which led to reduced pumping efficiency, clotting and the embolism.

Is it reasonable to expect a busy physician who is unfamiliar with the patient to read a long chart in order to ferret out this causal sequence in order to record it on a form? Clearly not, so an EHR connected FHIR app that did the 'research' for the physician could lead to more accurate death reporting and could even save the physician time. This was the premise for the FHIR app development project we will now discuss.

---

[5] https://palliative.stanford.edu/home-hospice-home-care-of-the-dying-patient/where-do-americans-die/

One of the clear advantages of FHIR is that apps have access to a patient's clinical data. They use the FHIR API to do this but are in no way limited to only that purpose and API. Hypothetically a web service could be developed based on the analysis of millions of patient deaths that a FHIR app could access in order to inform the physician of the likely causal sequence(s) of death in a specific patient based on the key clinical data it culls from the patient's record. This could be in the form of a timeline for the physician that highlights the historic pattern of clinical events that might have constituted the causal sequence of this patient's death. The physician can select from this sequence or add other events they feel were causal that were overlooked by the analytic engine.

This is precisely what the Death Worm (Death-on-FHIR) demonstration FHIR app aims to do.[6,7] The goals of the project are to:

- Integrate death reporting more seamlessly into physicians' workflows
- Save physicians time by pre-populating demographic and other information from the EHR
- Improve accuracy by providing physicians a visualization of decedent's medical history
- Assist physicians in making cause of death determinations by using an analytical engine to make predictions and recommend possible causal sequences based on information in the decedent's medical history
- Advance medical research and improve patient care by facilitating the exchange of coded mortality data and other population-level insights from electronic death registration systems to EHR systems

The initial phase of the project consisted of two major efforts. The first was mapping death certificate data elements to FHIR. The second was developing a clinical decision support tool for death reporting. As shown in Fig. 12.2, a proof-of-concept application was developed and it was tested against Cerner's FHIR developer sandbox to ensure compatibility. The State of Utah tested the app against Epic's sandbox in January 2018 at the Integrating the Healthcare Enterprise North American Connectathon in Cleveland.[8] At present, the tool includes visualizations of the decedent's health history but not yet recommendations on causal sequences. Doing this requires the third phase of the project which is developing an analytics engine that can suggest causal sequences based on the analysis of millions of death records.

Prior to 1992 the CDC was named the Centers for Disease Control. The word "prevention" was added in recognition of the growing public health problem pre-

---

[6] https://www.ncbi.nlm.nih.gov/pmc/articles/PMC5548492/

[7] https://github.com/nightingaleproject/nightingale

[8] http://www.iheusa.org/sites/iheusa/files/CAT18%20System%20Reg%20Kickoff%20Webinar_2017_08_23%20FINAL.pdf

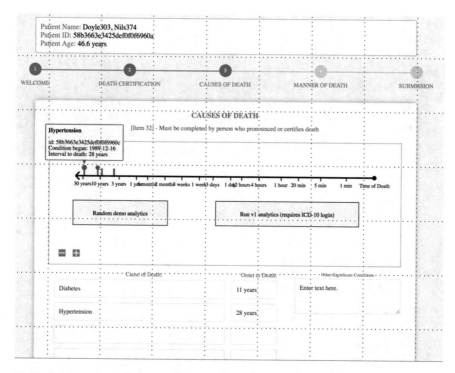

Fig. 12.2 The Death Worm demonstration FHIR app seeks to supply all the key clinical data from a deceased patient's record to an analytic web service that, based on analysis of millions of death certificates, in order provides the app with the information to visually present the possible causal death sequence to the physician who will be completing the death certificate. At present, the tool includes visualizations of the decedent's health history but not yet recommendations on causal sequences. (CDC)

sented by chronic disease and the frequency with behaviors are a root cause. This expansion of CDC's mission effectively expands the role of public health to encompass an understanding of the behaviors that lead to chronic disease and strategies for changing those behaviors. As a part of that it would be of obvious value if physicians could be made aware of their patients' behaviors outside of the limited purview they have when those patients are in the office.

Among the key behaviors of interest are diet and exercise. In the chapter on mHealth we discussed the technologies that now offer the promise of tracking patient activity and making that available to interested parties, including the physician. What about nutrition? We will now discuss a second demonstration FHIR app that may point the way toward an approach to providing physicians with a more complete view of their patient's nutritional status.

Demonstration FHIR APP: Improved Nutrition in Chronic Disease Patients

Cameron County is located in the Rio Grande Valley (RGV) region of Texas, adjacent to the United States – Mexico border. Based on the 2009–2013 Census it contains some of the most impoverished cities in the US. Its residents suffer from a high prevalence of chronic disease: approximately 28% incidence for diabetes, 49% for obesity, and 32% for hypertension in 2012.[9] The healthcare system in the RGV does not have sufficient resources to support these levels of chronic disease, so this project looked at novel strategies such as using informatics to encourage people to eat healthy foods. Healthy dietary and physical activity patterns can reduce chronic disease risk, reduce weight, and contribute to long-term health.[10] An extensive literature review suggested that behavioral dietary interventions have been successful in improving health outcomes, but are usually designed as short-term studies, rely on self-reported dietary intake, and do not engage healthcare providers.[11]

Given rapid adoption of FHIR by commercial EHR vendors, the project team felt it was timely to build a tool that integrates clinical data from the health environment with dietary data from outside the health environment to support new innovative dietary interventions.

The goal was to make it easy for clinics, hospitals, health information organizations, and public health agencies to work with grocery stores and food outlets to implement health promotion and diet monitoring programs. Given that participant-reported dietary data is subject to several biases, including recall bias and social desirability bias, the project selected more objective grocery purchase records as a surrogate measure for dietary habits. Several studies in the literature have demonstrated the utility of grocery purchase records to measure an individual's dietary habits.[12] However, to a great extent these records are kept in information silos at grocery stores, where they are most commonly used for marketing and business development. There are several

(continued)

---

[9] https://ncbi.nlm.nih.gov/pmc/articles/PMC5723109/
https://ncbi.nlm.nih.gov/pmc/articles/PMC4736910/
https://ncbi.nlm.nih.gov/pmc/articles/PMC3475522/#R10

[10] https://ods.od.nih.gov/pubs/2015_DGAC_Scientific_Report.pdf
https://ncbi.nlm.nih.gov/pubmed/27680992

[11] https://ncbi.nlm.nih.gov/pubmed/23510154
https://health.ny.gov/events/population_health_summit/docs/presentation_bringing_health_to_the_table.pdf
https://ncbi.nlm.nih.gov/pubmed/16390668

[12] https://ncbi.nlm.nih.gov/pmc/articles/PMC3243220/
https://ncbi.nlm.nih.gov/pmc/articles/PMC3900174/

efforts underway to create nutrition scoring algorithms based on data from grocery stores. These include NuVal, a front-of-package label scoring system based on the Overall Nutrition Quality Index nutrition algorithm developed by faculty at the Yale School of Public Health; NutriSavings, which uses a combination of "Ratio of Recommended to Restricted" and "Nutrient Rich Foods" to calculate nutrient density scores; and the Healthy Eating Index, produced by the USDA, which includes both aggregate and granular "component" nutrition scores.[13]

An initial proof-of-concept application was developed by a team of Georgia Tech students in spring 2017 working with Johnny Bender, at the time a student at the University Of Texas School Of Public Health in Brownsville. In his opinion it successfully demonstrated the integration of clinical data using FHIR with ongoing grocery purchase algorithm scores. The team conducted several contextual inquiries with nutritionists and registered dietitians in the RGV to identify the essential clinical items to include in the provider-facing application. These included weight (Observation BodyWeight), respiratory rate (Observation RespiratoryRate), height (Observation BodyHeight), glucose (Observation Glucose), hemoglobin A1c (Observation HbA1c), body mass index (Observation BMI), low-density lipoprotein (Observation LdlCholesterol), and high-density lipoprotein (Observation HdlCholesterol). To populate the grocery purchase algorithms, the team obtained the data sharing specification from one of the nutrition algorithm vendors and created a mock database of grocery purchase algorithms. Figure 12.3 presents screenshots from the final provider-facing application, which includes patient demographics, and tiles and trendlines with summarized vital signs, labs, and nutrition scores.

Future plans include expansion of the application by adding authentication support, enhancing the user interface for providers, adding a consumer user interface, adding support for direct grocery purchase data feeds from grocery stores, adding medication and diagnoses FHIR resources, adding EHR integration support using SMART on FHIR by deploying the application to the SMART application gallery, and adding support for a clinical decision support (CDS) Hooks workflow by creating a dietary habit informational card with SMART on FHIR application link, adding support for patient goals, and adding dietary recommendations based on historic purchases. When this enhanced solution is implemented in a real-world setting, the system will be able to track participant identifiers across grocery store and healthcare facilities by distributing grocery tracking cards with linked EMR record identification numbers to participants, who will scan the card at grocery point-of-sale systems during checkout.

(continued)

---

[13] https://ncbi.nlm.nih.gov/pubmed/26494178
   https://ncbi.nlm.nih.gov/pmc/articles/PMC3100735/
   https://ncbi.nlm.nih.gov/pmc/articles/PMC3810369/

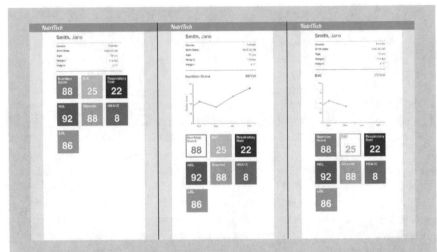

Fig. 12.3 Screenshots from the final provider-facing spring 2017 proof-of-concept nutrition score FHIR application. The first screen displays the relevant demographics, vital signs, labs, and nutrition algorithm score. Clicking on icons displays a trendline for the selected measure. The second screen displays the nutrition score trendline, and the third screen displays the body mass index (BMI) trendline

A key strategy to improve the management and prevention of chronic disease is to monitor preventive health measures, like physical activity and eating habits. These measures should be integrated into the clinical environment and used by innovative cross-industry partnerships, like those between healthcare institutions and grocery stores. Applications such as this one can support dietary intervention and monitoring programs, where nutritionists at grocery stores collaborate with providers at clinics and hospitals to make dietary recommendations to patients to improve their health.

12.5   The Future of Public Health

As we have discussed the foundation of public health is surveillance to identify risks and disease outbreaks and their causes in the population. A great deal of public health data comes from the healthcare system and it has historically been provided on paper and then mailed/faxed in to be put manually keyed into a database. This is a potentially error prone process and one that struggles to make data available in a timely manner. As the first case study illustrates, with EMRs now broadly in place, public health is moving to an electronic data collection process to improve accuracy of the data and timeliness of response.

A second mission of public health is to use the data it collects and analyses to feedback knowledge and advice to physicians as they deliver care. Increasingly this involves not just traditional data sources but data on the social determinants of disease development and outcomes. Today, as our second case study illustrates, this increasingly involves wearable devices and apps.

Epidemiology has always been foundational to public health going back at least to Snow's pioneering work on cholera in central London. Today, with a broader and deeper pool of "big data" to work with, overcoming data quality issues and providing context are challenges to be overcome.

Finally, a key objective that distinguishes much research from public health is the objective of using data to drive actions. Here again, as our case studies illustrate, the potential future importance of EHR connected or patient facing FHIR apps is hard to overestimate.

12.6 Population Health

We know that chronic diseases account for most US healthcare costs and that the design of our care delivery system is not well suited to chronic disease management. To correct this, we discussed the need for new care models better designed for the management of these diseases. We have also discussed the critical role that patient behaviors play in the development of chronic disease and in its successful management. Finally, provider reimbursement in the new care and payment models we discussed earlier in the course are usually based on care quality and clinical outcome metrics.

Population health began in response to the need of providers to understand and improve their performance against these metrics. It was defined by David Kindig MD, PhD, and Greg Stoddart, PhD in a 2003 paper in the *American Journal of Public Health* titled "What Is Population Health?" as "the health outcomes of a group of individuals, including the distribution of such outcomes within the group."[14] In that paper they argue that the field includes health outcomes, patterns of health determinants, and policies and interventions that link these two.

Like public health, it is an important use case for interoperability since individual providers in some network or system that has contracted to deliver cost effective outcomes will usually have many different EHRs. Finally, like virtually all other domains of health informatics, population health is increasingly employing analytic technologies to predict and proactively manage clinical results.

12.7 popHealth®

Today, given the growth in demand because of the increasing importance of value-based care, population health is a well-developed field dominated by commercial providers. In contrast, popHealth is an open-source web-based population health

---

[14] https://ajph.aphapublications.org/doi/abs/10.2105/AJPH.93.3.380

tool focused on the aggregation and reporting of the kind of clinical quality measures required to participate in a value-based contract. It includes a centralized repository of clinical data that is sent from EHRs via standards such as Consolidated Clinical Data Architecture (C-CDA) Continuity of Care (CCD) Documents or QRDA (Quality Reporting Data Architecture) Category 1 Documents.

Quality Reporting Document Architecture (QRDA) is an HL7 standard for quality reporting. QRDA reports are not patient-specific, protected clinical data but rather are de-identified statistics at one of three levels of detail, called QRDA categories:

- Category I (Patient-level) Reports: Data for one patient for one or more clinical quality measure
- Category II (Patient-list) Reports: Data for a set of patients for one or more clinical quality measures
- Category III (Aggregate-level) Reports: Aggregate data for one provider for one or more clinical quality measures

In popHealth, as shown in Fig. 12.4, Clinical Quality Measures (CQMs) are calculated for providers and presented with drill down ability to the provider and patient-level data. It allows providers to track trends in quality and health over time and calculate CQMs for Meaningful Use and other CQM purposes such as meeting the clinical quality reporting requirements of a value-based contract.

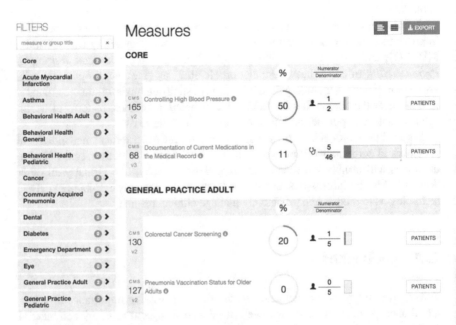

Fig. 12.4 popHealth's web interface presents aggregated clinical quality measures with drill down capability to provider and patient-level data. (Courtesy popHealth)

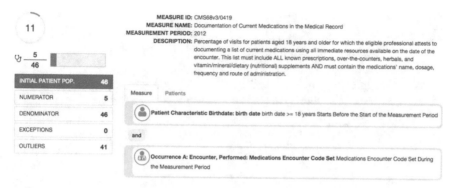

Fig. 12.5 The calculation of quality metrics requires a precise definition of the population of interest (the denominator) and the reportable subset of that population (the numerator). (Courtesy popHealth)

As shown in Fig. 12.5, the calculation of useful and accurate quality measures typically requires careful and potentially detailed specification of the population of interest (the denominator) and the reportable subset of that population (the numerator). The granularity of FHIR could be of importance in supporting this.

popHealth is an open source reference implementation, so anyone can modify and extend its functionalities and, under its license, is not required to contribute back fixes or innovations. This is nevertheless encouraged and appreciated and fosters mutual benefit for all parties.

Case Study: Wellcentive
With the growth in outcome-based contracting by insurance companies, Medicare and major employers, commercial population health management tools and systems are available for use by physician practices and health systems. The typical application is to measure and monitor the performance of a group of physicians who have contracted collectively to cost effectively provide an acceptable level of defined quality metrics. Wellcentive is one of the larger providers of these services, but certainly not the only one. Rather than repeat what we've already discussed, we will consider some features of their reporting that we've not yet seen. For example, a Wellcentive report of overall performance for a list of practice-defined alerts could include metrics specific to particular pay-for-performance or outcome-based contracts. These would require the collection and aggregation of data not specified under Meaningful Use.

(continued)

Figure 12.6 is an interesting example of Wellcentive's use of visual analytics and illustrates the use of a broader dataset than is traditional in quality reporting. The tool is written in JavaScript so it can run on a variety of computers, including mobile and tablet devices. It can be configured to show many metrics in multiple visualizations on the same dashboard or report. When this is done, they are "connected" so the user can highlight portions of one visualization and see the corresponding portion in each visualization, but with different metrics or dimensions.

In this case, we are seeing only one graphic that shows providers (each box is a provider) based on the percentage of their diabetic patients that are out of control (using one of the two general standards of an HbA1c of greater than 9%, the other standard is set at 7%) and by the percentage of diabetic patients in their practice but not seen in the prior year. The shading of each box correlates with the percentage of patients out of control, with darker indicating poorer performance. The size of each box correlates with the percentage of diabetic patients not seen in the past year, with a larger size indicating fewer patients with annual visits. Thus, the providers in the upper left corner who have the smallest and lightest boxes are, in the traditional sense, the top performers. They see their diabetic patients annually and have them under good control. However, the provider in the lower right corner is very interesting. This practice has a well-controlled diabetes patient population despite not regularly seeing their diabetic patients in the office. It would be interesting to compare this practice's care processes with those in the upper left corner. Perhaps, for example, the practice at the lower right is using remote monitoring and only sees patients who are getting out of control while not wasting time and money seeing those it knows to be in good control. This provider may, in fact, be the most cost effective and, hence, the most profitable practice under an outcome-based reimbursement model and their diabetic care process might be something the others should adopt. An analysis of this kind presents a more complex data aggregation and curation problem than is normally the case in quality reporting (which is typically based on a small and well-defined set of quality metrics) since it requires aggregating, normalizing and standardizing claims, EHR and clinical lab data.

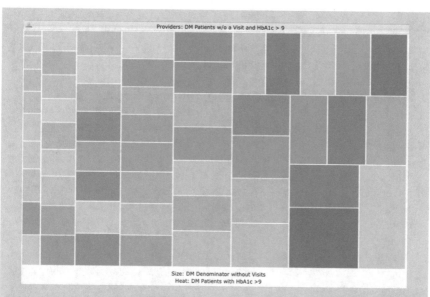

Fig. 12.6 A Wellcentive visual analytics report helps find the most cost-effective providers of diabetes care. Each provider is represented by a box with a darker box representing poorer control of that provider's diabetes population and a smaller box representing a higher percentage of annual visits for those patients. The providers at the upper left have the best control and also see their patients at least annually. The provider at the lower right has a virtually equal degree of control with far less frequent visits and may deliver the most cost-effective care. A further analysis of this provider's care process for diabetes would be worth doing to find approaches other, less cost-effective, providers might adopt. (Courtesy Wellcentive, All Rights Reserved)

12.8 A FHIR Solution for Population Level Diabetes Care

The innovative Wellcentive reporting tool we just discussed illustrates that companies in the population health field are seeking to add value to their products through analytic based tools. Rimidi uses analytics to support improved clinical care and is transitioning its technology from the one patient at a time paradigm to a population level care model.

Case Study: Rimidi
Rimidi is also an early example of a commercial enterprise based on FHIR and SMART on FHIR technology.[15] The company's platform technology enables personalized management of chronic diseases across a population. A key goal is to enable a continuum of care from the clinic to the home so the platform is used collaboratively between patients and their healthcare providers, a capability that

(continued)

---

[15] The author is an advisor to Rimidi and has received stock options as consideration for his efforts on behalf of the company.

is of increasing importance as we move from fee-for-service to value-based healthcare. As we have discussed, clinicians under value-based contracts need to be able to actively monitor a population of patients, identify problem areas, and leverage the right analytics to intervene in a meaningful way, hopefully before patients end up in the emergency department in need of hospital care. Similarly, patients need the tools and technology to understand how to manage their chronic health conditions yet also feel continuously supported by their healthcare team, not just during an appointment in the office.

Rimidi's cloud-based software interfaces directly to the electronic health record as well as to multiple wearables and medical devices used by patients to aggregate relevant data for specific chronic diseases such as diabetes and heart failure. The company says that SMART on FHIR facilitates rapid integration across multiple health IT systems. As shown in Fig. 12.7, using Cerner's SMART on FHIR environment, Rimidi is able to dynamically create patient and healthcare provider accounts within the provider's existing clinical workflow in order to provide a seamless end user experience without extra authentication steps or dual data entry. Rimidi also integrates to consumer devices via API and Bluetooth standards as shown in Fig. 12.7. The company feels that this ease of integration is key to end user acceptance.

The clinical and patient generated health data are also used to drive risk analysis and triage patient care for care managers. The Rimidi system identifies

Fig. 12.7 The Rimidi application shown here runs within a SMART window in the same area of the Cerner Millennium EHR that physicians use to chart and review other data. The application is initiated from the same left menu that clinicians use to initiate other EHR functions. The Wellbeing Score represents how an individual is meeting standards of care overall and is based on practice guidelines for comprehensive diabetes care from the American Diabetes Association, American Heart Association and Joint National Committee. Note that this diabetes specific display combines clinical data obtained from the EHR using FHIR with data recorded at home by the patient thus providing the physician with a more comprehensive view of their patient. (Courtesy Rimidi)

(continued)

patients that are not receiving a standard of care and highlights those individuals for the care team to review.

The platform also provides predictive modeling capabilities for clinicians caring for the patient, as shown in Fig. 12.8. This is done by a clinical decision support module that allows clinicians to visualize the predicted impact of treatment decisions (such as insulin dosage adjustment) on patient outcomes (such as blood glucose levels during the day). The company feels that this predictive capability shortens the time to achieve optimal management of patients and, thereby, increases clinical capacity.

As shown in Fig. 12.9, Rimidi is adding the ability to identify patients meeting criteria for new and emerging treatment options such as novel medical devices, new pharmaceutical indications, or updates to practice guidelines. The goal is to shorten the adoption cycle as new care pathways, devices, or pharmaceuticals demonstrate patient benefit.

While a dashboard can conveniently indicate which patients might qualify for new treatments, the company hopes that notifying clinicians within their workflow will further help overcome clinical inertia and speed adoption. To implement this goal Rimidi offers integration to third party SMART on FHIR applications running on its platform which is device agnostic and serves as a

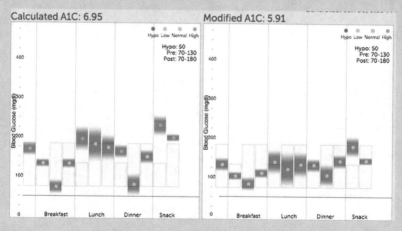

Fig. 12.8 A predictive model of patient blood glucose control by time of day (a modal day distribution) showing before and after visualizations based on simulated insulin dose adjustments. The open bars indicated target glucose levels throughout the day. The dots are predicted levels and their colors correspond to their relationship to the target level. (Courtesy Rimidi)

(continued)

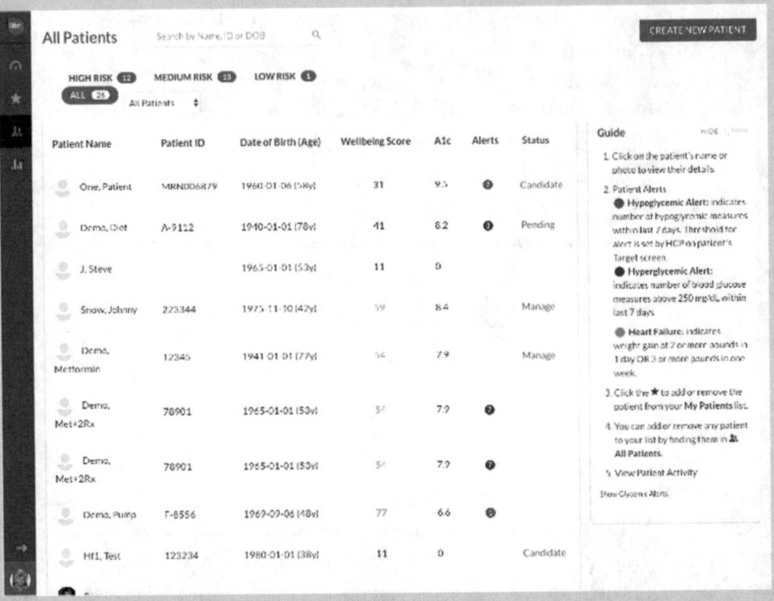

Fig. 12.9 The provider dashboard displays patient identifiers, wellbeing scores, most recent hemoglobin A1c's and blood glucose alerts. The wellbeing score is used to triage patients into high, medium or low risk categories. Patients identified for new treatments based on patient characteristics and treatment criteria are also identified on the dashboard. (Courtesy Rimidi)

middleware layer between the EHR and patient-facing apps and devices. For example, both continuous glucose monitoring (CGM)[16] and flash glucose monitoring (FGM)[17] are areas of rapid innovation and growth. New devices are coming to market and gaining reimbursement for greater segments of the diabetes population. CGM was recently approved for Type 2 diabetes by CMS. Rimidi's platform can identify patients that qualify for CGM such that clinicians can best match care pathways and new technologies to patient needs. The company feels that this approach will enable flexibility and adaptability of treatment management as new innovations come to the consumer market.

[16] https://www.fda.gov/newsevents/newsroom/pressannouncements/ucm534056.htm
[17] https://www.fda.gov/NewsEvents/Newsroom/PressAnnouncements/ucm577890.htm

12.9   FHIR Bulk Data Protocol

Earlier we discussed the provisions of the 21st Century Cures Act with respect to the provision of APIs by certified EHRs. Another aspect of the act is its effect on health information exchange for purposes such as population health. In January, 2018 ONC released its "Draft Trusted Exchange Framework and Common Agreement"[18] Interestingly, it extends to patients and includes data from consumer driven devices. It also develops a mechanism to connect Health Integrated Networks ("Qualified HINs") across the country.

Given the rising importance of value-based care and the resulting need for providers to submit quality metrics for large groups of patients, it is not surprising to see a focus on API-based data access: "Within twelve (12) months after the FHIR standard with respect to Population Level Query/Pulls has been formally approved by HL7, each Qualified HIN shall cause its Broker to be able to initiate and respond to all Query/Pulls for as many individuals as may be requested by another Qualified HIN in a single Query/Pull."

In other words, once HL7 extends the standard to support an emerging "Bulk Data Access" protocol, ONC wants to see all Qualified HINs support it. While FHIR allows for access to data from multiple patients at a time, at present the focus of actually applying the standard (e.g. Argonaut) is primarily aimed at the support of direct patient care, one patient at a time. Moreover, the current FHIR REST API is an inefficient means of querying the records of a large group of patients, since it relies on a client to follow "next page" links one at a time in order to fetch a complete data set.

The new protocol would provide an important extension to the FHIR standard and one that may have far reaching consequences for not only public and population health but for quality reporting and research. For example, as we will discuss later, machine learning is an increasingly important approach to abstracting 'knowledge' from the records of many patients. This protocol could facilitate bulk access to the needed data.

HL7 working together with SMART Health IT and the FHIR community have posted a Bulk Data API proposal on the HL7 site[19] and Josh Mandel and Dan Gottlieb have posted a presentation on the API.[20] Grahame Grieve has also posted a discussion of the protocol on his blog.[21]

Two example API queries are shown below. The first returns all data on a specified patient, and the second returns all data on a specified group of patients:

---

[18] https://www.healthit.gov/sites/default/files/draft-trusted-exchange-framework.pdf

[19] http://hl7.org/fhir/2018Jan/http.html#async

[20] https://docs.google.com/presentation/d/1QpMUIohFEJJcxxrcWfx80sRt XDWFIQBCIKw1BXsy454/present?slide=id.p

[21] http://www.healthintersections.com.au/?p=2689

```
[
{"id": "06eb35fc-09e6-48 ... "given": ["Lucille"],"family": "Bluth"}]} ,
{"id": "cf53f382-6eb6-4f ... "given": ["George", "Oscar"],"family": "Bluth","suffix": ["Senior"]}]} ,
{"id": "406a9c3e-50f9-4c ... "given": ["Michael"],"family": "Bluth"}]}

]
```

Fig. 12.10 An illustration of an ndjson formatted file that might be created as a result of a bulk data query. Inspection of this should reveal that this file is functionally the JSON equivalent of the ubiquitous CSV file format. It can be converted to the CSV format using free online tools. Like a CSV file, it is easily parsed for whatever the intended purpose might be. (Courtesy Josh Mandel)

```
GET [base]/Patient/$export?start=[date-time]&_type=[type,type]

GET [base]/Group/[id]/$export?start=[date-time]&_type=[type,type]
```

The definition of a group is not yet clear but existing FHIR servers provide searching capability that can return a group of patients based on the user's specification. At present the decision as to what criteria to support is server specific but this might become a part of the Bulk Data Protocol. For now, the main focus is providing a mechanism to get all patients or a specified group of patients (such as those with a particular disease or with specified insurance coverage) into a file for use in a system that, unlike an EHR, is optimized for doing analytics without impacting response time to clinical queries for patient care.

A bulk data request could take quite some time to process so there is also the concept of an asynchronous protocol in which the FHIR service might receive the request, process it enough to ensure it is valid and then return an HTTP accepted (202) response along with a URL where a manifest of data files will eventually be available. During the processing the requesting entity could obtain an indication of progress and an HTTP success (200) message when it is complete. The data is returned as a set of files in Newline delimited JSON (ndjson), a format for storing or streaming structured data that may then be processed one record at a time (Fig. 12.10).[22]

While the ONC request includes the concept of a subscription in which data is automatically sent out as it becomes available, Grahame's post suggests that this is too demanding of the servers and rests on the idea that requests can be made as frequently as needed.

In general machines would be executing queries, and they would not have to perform a full export every time. Instead, a syntax is being discussed so they could ask for "only things that have changed since $timestamp". All of this is subject to change, of course, since the current description is very preliminary but even at this early stage it is important because of the potential impact of this new FHIR capability on public and population health, quality reporting and research.

---

[22] http://ndjson.org/

For technically skilled readers there is a detailed discussion of the Bulk Data Proposal posted on GitHub.[23] There is also an interactive demonstration posted on the SMART site.[24]

12.10   Recap

We have seen that public health is the oldest domain in which statistics and eventually informatics have been applied in healthcare. Population health is one of the newest and is a clear example of the impact that economic factors have on the demand for informatics systems and tools.

Both of these domains rest heavily on interoperability since they involve the aggregation and analysis of large cohorts of clinical and other data. Both also present clear opportunities for innovation using FHIR as to how they aggregate and use that data. In this chapter we have seen a few early examples of that innovation and it is highly likely that there will be a great deal more of it over the coming years.

---

[23] https://github.com/smart-on-fhir/fhir-bulk-data-docs

[24] https://bulk-data.smarthealthit.org

Chapter 13
Analytics and Visualization

13.1 Health Data at Scale

This book is not about how to do analytics. Rather, we will explore the specialized infrastructures and tools that support research on and exploration of health data at scale. Generally, this involves aggregating at least the clinical part of that data from a large number of digital health records. This introduces the interoperability issues and technologies we have been discussing throughout the book. It is good to think back to the proposed FHIR Bulk Data Protocol we discussed in the population health section. We referred then to research as another future application of FHIR once that access method becomes available.

Advanced analysis of these big healthcare datasets offers the promise of more precise, personalized care and the use of predictive tools to anticipate and treat clinical problems much earlier than was possible in the past. This analysis must overcome many typical problems and limitations of EHR data. Despite the challenges, we are clearly heading into a new era of analytics-powered medicine and healthcare. Hopefully, this chapter provides you with at least a feel for what that might look like.

13.2 Big Data

The world is awash in data. It is growing at exponential rates. People have coined the term "big data" to refer to this phenomenon but often cannot agree on what it means. What separates big data from everything else? Cesar Hidalgo at MIT's Media Lab says that, to qualify for this distinction, data must be big in size, resolution, and scope. To reframe this idea in a way that is directly relevant to transforming healthcare delivery systems:

© Springer International Publishing AG, part of Springer Nature 2018
M. L. Braunstein, *Health Informatics on FHIR: How HL7's New API
is Transforming Healthcare*, https://doi.org/10.1007/978-3-319-93414-3_13

- Data must represent many patients and providers;
- It must do so in detail, and, as we will now discuss;
- It must be combined with other data to give the real world context within which patients live and get sick, care is delivered and the delivery system operates.

13.3   Real World Data

Real-world data (RWD) is a relatively new term encompassing clinical data derived from a number of sources that affect clinical outcomes in a heterogeneous patient population in real-world settings. These include

- Surveys and population level data
- EHR data
- Observational studies
- Clinical trials
- Registries

My Georgia Tech colleague Dr. Jon Duke and his collaborators, who developed the OMOP Common Data Model[1] we will explore later, use the term 'observational data' for essentially the same thing. We also referred to this kind of data in our discussions of public and population health so their impact goes far beyond individual patient care.

The hope is that analysis of this richer and broader dataset will generate meaningful insights into unmet needs and interventional strategies with greater clinical and economic impact on patients and healthcare systems. There are of course new challenges introduced when the range and scope of data sources expand. These include:

- Privacy and security
- Diversity across geography and stakeholders
- Statistical rigor
- Cost of collecting and maintaining such large datasets

To understand the potential value of real world data, read a short but interesting and insightful article by a trio of researchers at Harvard's Center for Biomedical Informatics. It begins with this paragraph "It has been argued that big data will enable efficiencies and accountability in health care. However, to date, other industries have been far more successful at obtaining value from large-scale integration and analysis of heterogeneous data sources. What these industries have figured out is that big data becomes transformative when disparate data sets can be linked at the individual person level. In contrast, big biomedical data are scattered across institutions and intentionally isolated to protect patient privacy. Both technical and social

---

[1] https://www.ohdsi.org/data-standardization/the-common-data-model/

challenges to linking these data must be addressed before big biomedical data can have their full influence on health care."

The paper then goes on to discuss and address this linkage challenge. The authors point out that, while electronic health records (EHRs) provide depth by including multiple types of data (e.g. images, notes, etc.) about individual patient encounters; claims data provides longitudinality – a less clinically detailed but more continuous view of a patient's medical history over an extended time period. Medical data may be episodic and incomplete if a patient receives care from multiple providers. Typically, there is only one payer for a time-period so that payer's claims from within that period should be more a complete record of care delivered. On the other hand, claims data typically lacks some important clinical details. A request for payment for a lab test order would be in a claim but the results would not be included, as they are in an EHR.

As a result, the authors of this paper say "Linking data adds value when they help fill in the gaps. With this in mind, it becomes easier to see how nontraditional sources of biomedical data outside of the health care system fit into the picture. Social media, credit card purchases, census records, and numerous other types of data, despite varying degrees of quality, can help assemble a holistic view of a patient, and, in particular, shed light on social and environmental factors that may be influencing health."[2]

13.4   Data Aggregation

Obtaining and aggregating health data presents many domain specific challenges. These include HIPAA and the protection of patient privacy as well as obtaining patient permissions for the use of their data unless it is fully de-identified. However, depending on the use case, full de-identification can substantially reduce the value of the data. Where genomic data is involved, de-identification may not be feasible if that data is to have any value.

Even with patient permissions, all the interoperability issues we discussed earlier can be impediments to aggregating the data. These include the reluctance of hospitals and other sources of clinical data to lose control of it. Without semantic interoperability, and it is still rare, data from multiple sources must often be normalized to a standard model to be useful for analysis. Earlier, we discussed the federated research model and the use of the Observational Medical Outcomes Partnership (OMOP) data model across the global OHDSI research network to solve many of these issues. We will now discuss some examples of research that takes advantage of OHDSI's global reach.

---

[2] http://jamanetwork.com/journals/jama/fullarticle/1883026

13.5 Research Using the OHDSI Network

We previously introduced the OHDSI federated research network. The vision of its developers was to create a tool for accessing the clinical experience of hundreds of millions of patients across the globe. Today OHDSI researchers can leverage data for more than 680 million patients from 56 databases in 12 countries.[3] All of this data has been mapped by the OHDSI collaborators into the Observational Medical Outcomes Partnership (OMOP) Common Data Model (CDM) format. We also discussed OHDSI's suite of applications and data exploration tools.

Figure 13.1 shows the coding systems used in OMOP and provides a good high level view of the data types it contains. Note the medication information is the largest data category which is not surprising since research around the use of medications was a primary use case for OHDSI and medication coding systems such as NDC often have tens of thousands of concepts.

ATLAS is one of the OHDSI tools and is used for exploration, standardized vocabulary browsing, cohort definition, and population-level analysis. Figure 13.2 shows an example of using ATLAS for cohort definition, an essential early step in research. What is unusual is that a cohort definition created using ATLAS can be shared, interpreted and executed on a global scale.

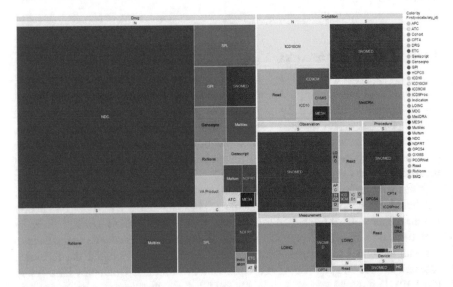

Fig. 13.1 OMOP uses many existing data standards. This illustration groups them by domain (e.g. drugs, conditions, procedures); by standard class and by the specific standard vocabulary (https://www.ohdsi.org/wp-content/uploads/2014/07/2017-08-OHDSI-medinfo-panel-Huser-vLOCAL-1356-noEXTRAs.pptx.pdf). Medication codes are the largest category because of the initial OHDSI use case and the many thousands of pharmaceutical agents in use. (Courtesy Dr. Jon Duke)

---

[3] https://www.ncbi.nlm.nih.gov/pmc/articles/PMC4815923/

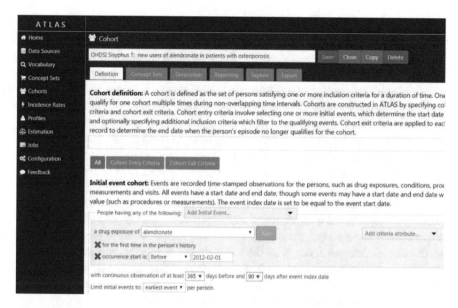

Fig. 13.2 Cohort definition using ATLAS. This example is specifying patients who took alendronate (a drug used to prevent and treat certain types of bone loss (osteoporosis)) for the first time prior to 2/1/2012 and where clinical data is available for at least a year before and 90 days after the initiation of therapy (https://www.ohdsi.org/wp-content/uploads/2014/07/2017-08-OHDSI-medinfo-panel-Huser-vLOCAL-1356-noEXTRAs.pptx.pdf). (Courtesy OHDSI)

13.6   OHDSI Use Case 1: Risk of a Serious Drug Side Effect

This first use case demonstrates an example of an originally anticipated use case for OHDSI. A series of individual patient case reports suggested that there might be an increased risk of severe facial swelling (angioedema) in patients taking the anti-epilepsy medication levetiracetam. Angioedema is rare so a study of its incidence requires access to a large number of patients. Such a study would be expensive and hard to execute in a timely manner using traditional methods.

An OHDSI study across 10 databases found nearly 75,000 seizure patients exposed to the drug and compared them to an equal number of patients on phenytoin, another anti-epilepsy drug. Patients were required to have an exposure to levetiracetam or phenytoin as well as a diagnosis of seizure disorder (as determined from a list of codes using in OMOP) prior to the exposure date. Patients were also required to have clinical observations for at least 6 months prior to drug exposure and could not have had a previous diagnosis of angioedema. Angioedema events were rare in both groups and the risk for levetiracetam was similar or lower than for phenytoin.[4]

The ability to conduct a study at this scale in a timely (3 months from conceptualization through data analysis) and inexpensive manner is remarkable. Its success suggests the transformative potential of digital health data when it is stored within an interoperable research framework.

---

[4] http://onlinelibrary.wiley.com/doi/10.1111/epi.13828/full

13.7   OHDSI Use Case 2: Care Pathways

This use case illustrates a more novel use of OHDSI. The consistent use of the best medical evidence in patient care is the key goal of a Learning Health System. This research was done using 11 OHDSI data sources from four countries, including electronic health records and administrative claims data on 250 million patients. The goal was elucidating clinical treatment pathways for type 2 diabetes mellitus, hypertension, and depression and comparing them to understand variations in treatment patterns (e.g. in the use of best medical evidence) around the world.

Providers treating diabetes were much more consistent in their choice of initial medication (metformin) than when treating hypertension or depression. Overall the variation in treatment patterns was wide for all disease areas, with entirely unique pathways found in 10% of diabetes and depression patients and almost 25% of hypertension patients. Results were similar using either claims or electronic health record data. While our focus is on the validation of federated research networks at scale it is worth noting that the study authors caution that the disease specific variation in the patterns of care they observed are "perhaps related to the availability, appropriateness, or acceptance of concrete recommendations".

A key observation made by the researchers in this study about the OHDSI technology is that it validates the core approach by showing that "a query authored at one OHDSI site may be run at all sites without further modification."[5]

13.8   Different Questions and Methods

Healthcare is a diverse and rich field for research and analytics. We will now consider a few examples to explore that diversity. The historic gold standard for developing new medical knowledge is the controlled clinical trial. Give two randomly selected patient groups alternate therapies (or a new therapy and a placebo) and see which group does better over time. These are difficult to do. Finding and recruiting the right patients is hard. These trials take a long time and they are expensive. However, if the question is which treatment works best in actual human patients, there is currently not a good alternative approach.

Now consider the problem of early diagnosis of a disease. Congestive heart failure (CHF) occurs when the heart muscle does not pump blood as well as it should. It affects 5.7 million people in the United States and there are over 200,000 new cases each year. A 2013 paper calculates that the total direct cost of heart failure is from $60 to 115 billion annually.[6]

---

[5] http://www.pnas.org/content/113/27/7329.full#ref-17
[6] http://onlinelibrary.wiley.com/doi/10.1002/clc.22260/pdf

According to a review article in Mayo Clinical Proceedings, "No single test can be used to establish the clinical diagnosis of heart failure. Instead, history and physical examination findings showing signs and symptoms of congestion and/or end-organ hypo-perfusion are used to make the diagnosis. Imaging studies documenting systolic or diastolic dysfunction and biomarkers are helpful adjuncts."[7] Thus, the diagnosis may require many types of clinical data. Using these data to answer this type of question is the domain of predictive analytics.

Finally, a different question. What is the optimal treatment strategy among already known and available options for a given group of patients? As opposed to the classic research questions, determining optimal treatment would require many alternative experiments, that is, trying every possible treatment strategy on groups of similar patients. These types of questions are best addressed through modeling and simulation using digital health data. There are a number of techniques for doing this.

13.9   Children with Complex Chronic Conditions

Children with complex chronic conditions accounted for 10.1% of pediatric admissions in 2006. These admissions used around 25% of pediatric hospital days, accounted for some 40% of pediatric hospital charges, caused over 40% of pediatric deaths, and used 75% or more of technology-assistance procedures. Given the seriousness and cost of care for these children, understanding them and predicting changes in their condition is of great interest. We will now describe work done by Georgia Tech's Dr. Jimeng Sun in collaboration with Children's Healthcare of Atlanta (CHOA) to do just that.

Figure 13.3 is a simplified graphic depicting the series of steps used by Dr. Sun in this work.[8]

First, we will focus on the Feature Construction phase. You know that the commonly used data standards for diagnosis, procedure and medication coding are very complex. Clinical Classifications Software (CCS) is a tool developed and continuously updated by the Agency for Healthcare Research and Quality (AHRQ) for clustering patient diagnoses and procedures into a manageable number of clinically meaningful categories.[9] For example, ICD-10 CCS aggregates illnesses and conditions into 259 mutually exclusive categories, most of which are clinically homogeneous. In this research 73 of the CCS features were found in the feature construction and selection phases to have sufficient predictive power to be included in the final model for predicting status change of the medical complexity of the pediatric patients (CHOA).

As listed in the Prediction box, various approaches to prediction (classification models) were tried and two – Random Forest (RF) and Gradient Boost Decision

---

[7] https://www.ncbi.nlm.nih.gov/pmc/articles/PMC2813829/

[8] https://amia2017.zerista.com/event/member/389218

[9] https://www.hcup-us.ahrq.gov/toolssoftware/ccs/ccs.jsp

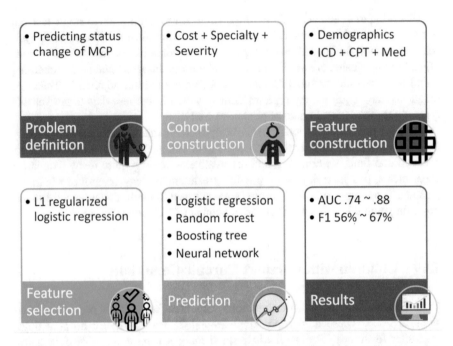

Fig. 13.3 An analytics pipeline (series of steps) used by Dr. Jimeng Sun's lab at Georgia Tech to predict changes in children with complex chronic conditions. It proscribes a logical series of steps that begins with problem definition and concludes with results. (Courtesy Dr. Jimeng Sun)

Tree (GBDT) – produced the best results. RF depends on the random selection of both data and independent variables to create many decision trees. The output is the mode or mean prediction of the individual trees. Gradient boosting is a machine learning technique that produces a prediction model using regression on an ensemble of weak prediction models, typically decision trees.

To understand the evaluation of models it is important to recognize that the models are attempting to classify a group under study. You may recall that earlier we discussed a model to predict which patients would likely develop sepsis over some defined future time period. This is a binary classification problem to divide a group of patients into two subgroups at some threshold of confidence based on a model. In this case it is children with complex chronic conditions who may or may not be likely to have a change of condition.

A binary predictive model can produce True Positives (TP) (in this case children predicted to have a change in condition who in fact do have one), False Positives (FP) (in this case children predicted to have a change in condition who do not have one) as well as True Negatives (TN) and False Negatives (FN).

Models are evaluated based on their accuracy: the ratio of correct predictions to total predictions [(TP+TN)/(TP+FP+FN+TN)]; their precision: the ratio of correct positive predictions to total positive predictions [TP/(TP+FP)]; and their recall: the ratio of correct positive predictions to the entire class of those who actually had the result being predicted (i.e., status change of medical complexity) [TP/(TP+FN)].

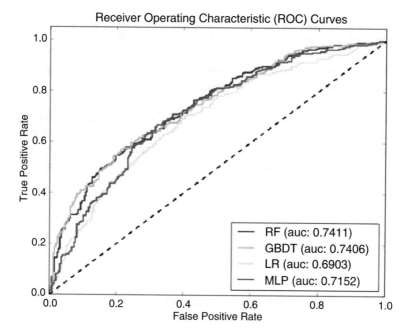

Fig. 13.4 This set of ROC curves compare the predictive power of models using RF: Random Forest, GBDT: Gradient boosting decision tree, LR: Logistic regression and MLP: Multilayer perceptron (i.e., a feedforward neural network with one hidden layer) techniques. (Courtesy Dr. Jimeng Sun)

The Results box in the pipeline reports two metrics: AUC and F1 Score. The F1 Score is the weighted average of precision and recall. AUC is the 'area under the curve' of the Receiver Operating Characteristic (ROC) curve, an example of which is shown in Fig. 13.4. This particular example compares the predictive power of models using RF: Random Forest, GBDT: Gradient boosting decision tree, LR: Logistic regression and MLP: Multilayer perceptron (i.e., a feedforward neural network with one hidden layer) techniques.

The ROC curve illustrates the diagnostic ability of a binary classifier system as its discrimination threshold is varied. Each point plots the true positive rate (TPR) against the false positive rate (FPR) at various threshold settings. In that graph each point represents a sensitivity/specificity pair corresponding to a particular decision threshold. The area under the curve (AUC) is equal to the probability that a classifier will rank a randomly chosen positive instance higher than a randomly chosen negative one.

13.10   i2b2 Introduction

Informatics for Integrating Biology and the Bedside (i2b2) is an NIH-funded effort based at Partners HealthCare System in Boston. Its mission is to enable clinical investigators to conduct research using state-of-the-art genomics and biomedical informatics. i2b2 supports complex research environments but provides an easily understood database schema that can be used to create a data warehouse where large clinical datasets abstracted from an enterprise EHR can be stored for future analysis. This is important because the database schema of commercial enterprise EHRs can be dauntingly complex, proprietary and not necessarily designed well to support ad hoc queries.

i2b2 implementations are comprised of "cells" that communicate via web services and, in aggregate, form an i2b2 "hive". The role of most of these cells should be clear from their names. Examples include de-identification of data, natural language processing and annotating imaging data. Custom cells can be added to the hive and there is an organized i2b2 users group to share information and applications and to create the potential for research collaboration and federated queries across institutions. A FHIR cell has been developed to support SMART on FHIR apps running against the data in an i2b2 instance. There is an interactive web client you can use to see how it operates.[10]

13.11   i2b2 Based Federated Research

The National Institutes of Health's Clinical and Translational Science Awards (CTSA) program seeks to facilitate the translation of research into clinical practice among a group of collaborating institutions. To facilitate this mission, Harvard Medical School created SHRINE (Shared Health Research Informatics NEtwork), a web-based software network. SHRINE links the respective i2b2 instances of five of Harvard's CTSA partner hospitals for the sharing of aggregated counts of patients meeting selected inclusion and exclusion criteria for demographics, diagnoses, medications, and labs. The goal is for SHRINE to enable population-based research, assessment of potential clinical trials cohorts, and hypothesis formation for follow-up study by combining the EHR assets across the hospital systems.

The Department of Biomedical Informatics at Harvard Medical School is working with healthcare centers across the U.S. to develop the Scalable Collaborative Infrastructure for a Learning Health System (SCILHS, pronounced "skills").[11] Each site will install a software suite and comprised of:

- i2b2, creating a federated query and response system using SHRINE
- The SMART on FHIR platform
- The Indivo PHR
- The REDCap (Research Electronic Data Capture) tool used to survey patients online

---

[10] https://www.i2b2.org/webclient/

[11] https://www.ncbi.nlm.nih.gov/pmc/articles/PMC4078286/

Demonstration FHIR App: Analytics-based Clinical Decision Support for Optimal Hospital Discharge

In spring, 2016 team of Georgia Tech graduate students developed a prototype of an analytics-based Clinical Decision Support FHIR app to help surgeons make optimal hospital discharge decisions. The project was done under the direction of Dr. John Sweeney, Joseph Brown Whitehead Professor and Chair, Department of Surgery, Emory University School of Medicine and Dr. James C Cox, Noah Langdale Jr. Chair in Economics, Georgia State University.

Through this project we will see the role that innovative analytic models can play in bringing new tools to the delivery of patient care. This app is also an excellent example of an informatics tool that is now of interest when hospitals are paid for delivering cost effective results and might be of less (or no) interest if they were still paid under a pay for procedure model.

The decision to discharge a patient from the hospital is a central determinant in the length of hospital stay (LOS) and of the risk of readmission to the hospital within 30-days of discharge. The behavioral tendency by healthcare providers is often to minimize the risk of readmission by keeping a patient in the hospital longer, driving up the length of stay and thereby sometimes unnecessarily driving up costs and depriving the hospital of the potential revenue from being able to handle more cases with its available bed capacity. On the other hand, a premature discharge can increase the risk of readmission leading to lower quality of care and penalties for health systems. For example, Medicare typically will not pay for a readmission for a clinical problem within 30 days of discharge for that problem.

Physicians and surgeons now make discharge decisions based mainly on subjective evaluations of available information on patients' status. As a result there is considerable variation in these decisions and it is therefore unsurprising that close to 20% of Medicare patients are readmitted within 30 days of discharge (a sign of low quality) which results in additional hospital charges totaling about $18 billion. Under current healthcare policy, physicians and hospitals are facing increasing pressure to reduce costs of hospital stay and reduce unplanned readmissions.

The app is based on a model which was developed by analyzing 30,000 de-identified patient records of general surgery patients. Recommendation on the readiness of a patient to be discharged from the hospital is based on the predicted class probabilities and historical target rates of readmissions.

To develop the model, a Random Forest (RF) model was trained and used to derive daily probabilities of patients being classified as ready for discharge while in hospital. The predicted discharge probability class is added in the set of features for readmissions. Next, a second RF model was trained to capture the success of discharge. Daily predicted class (discharge or not, readmit or not) probabilities are derived from both models and are provided to physi-

(continued)

cians at the point of care using the friendly visualizations shown in Figs. 13.4 and 13.5 and accompanied by the ranking of features in terms of importance.

The user interface was developed using principles from behavioral economics to increase its effectiveness and acceptance. Its goal is to advise surgeons to discharge patients at the earliest time when the model suggests this is the proper decision in order to minimize both length of stay and the readmission risk. Very importantly, physicians are unlikely to believe a 'black box' so the model and the app are designed to make it clear why a discharge recommendation is being made at a particular date and time.

To accomplish this the SMART on FHIR app visually provides the details behind the recommendation and a choice between rectangular block or circular display of the estimated marginal effects of patient variables on readmission probability, as shown in Figs. 13.5 and 13.6. In both, the recommended action of Do Not Discharge, Physician Judgment, or Discharge (DISCHARGE in this case) is provided at the upper right. Three-color coding in the middle section brings attention to patient variables that are (a) within normal range (green) or (b) marginally outside normal range (yellow) or (c) significantly outside normal range (red). The size of each variable area indicates its marginal effect on the recommended action. Selecting any patient variable in the variable information display opens a graphical display (on the right) of the time series of that variable from admission to the current day.

Also, in both displays, the left part of the screen displays estimated readmission probabilities over time and 80% confidence intervals, together with target readmission probability for the procedure code for all hospital stay days up to the current day. The recommendation is Do Not Discharge whenever the estimated probability of readmission exceeds the target rate. It is physician judgment whenever the target readmission probability falls between the point estimate of readmission probability and the upper bound of the confidence interval. The recommendation is Discharge whenever the estimated upper bound for readmission probability falls below the target readmission probability.

This app was a good early use case for FHIR since the data that the app needs to operate consists mostly lab test results and clinical observations that are readily available from any FHIR-enabled EHR in the form of FHIR Observation resources. Those resources will be similar (but not identical due to differences such as in coding) no matter the underlying EHR. This means that one app can run against any FHIR enabled EHR with a manageable amount of 'tweaking' greatly simplifying development and maintenance of the app. The SMART on FHIR app platform, if properly integrated into the EHR, makes the app appear to the physician almost as if it were an integral part of the EHR, greatly facilitating workflow and process.

(continued)

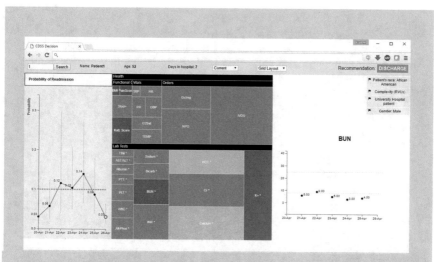

Fig. 13.5 The rectangular block display consists of three areas. (Left to Right) (1) A time series graph of predicted readmission probabilities; (2) a Block Display of current marginal effects (sizes shown by relative surface areas) of clinical variables on readmission probability with color coding of each variable being determined by whether the value is normal (green), moderately abnormal (yellow) or critically abnormal (red); (3) an example display of BUN time series data (normal values within the dotted lines); and the Discharge Recommendation (top right)

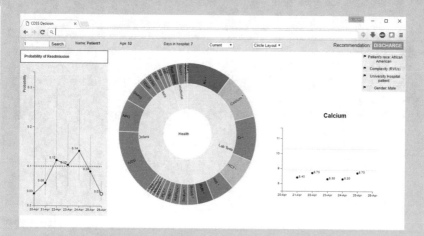

Fig. 13.6 The circular display consists of three areas (Left to Right) (1) A time series graph of predicted readmission probabilities; (2) a circular display of current marginal effects (sizes shown by relative surface areas) of clinical variables on readmission probability with color coding of each variable being determined by whether the value is normal (green), moderately abnormal (yellow) or critically abnormal (red); (3) an example display of Calcium time series data (normal values within the dotted lines); and the Discharge Recommendation (top right)

Case Study: vRad Radiology

vRad, founded in 2001, is now part of MEDNAX[12] and is the largest provider of tele-radiology services with over 500 board certified radiologists reading some 20,000 cases daily for its 2,100 hospital clients in the US using the company's patented picture archiving and communication system (PACS) and broad telemedicine platform. The company's services are heavily based on informatics. We will focus here on vRad's use of natural language processing (NLP) and deep learning to analyze medical images. We will focus on stroke to see how deeply these technologies are embedded into the company's workflow and processes and the clinical importance of the more timely response their use can provide.

First we will consider NLP. There are two subcategories of stroke that require very different treatment. Ischemic (thrombotic) stroke is caused by a blockage of the blood flow to an area of the brain. Intracerebral hemorrhage is bleeding from a rupture of the arteries that supply blood to the brain. According to the 2015 guidelines "Timely restoration of blood flow in ischemic stroke patients is effective in reducing long-term morbidity. For patients who meet national and international eligibility guidelines, intravenous recombinant tissue-type plasminogen activator (r-tPA) improves functional outcomes at 3–6 months when given within 4.5 h of ischemic stroke onset and should be administered. Every effort should be made to shorten any delays in the initiation of treatment because earlier treatments are associated with increased benefits."[13]

However, the presence of active bleeding or a problem with blood clotting is a contraindication to the administration of IV rtPA for the treatment of acute ischemic stroke.[14] Imaging studies, such as a CT scan, are "the initial study of choice for evaluating an acute stroke patient - it is used for inclusion criteria and to rule out hemorrhage."[15]

Clearly the timely availability of the results of an imaging study are critical to accurate and effective treatment of stroke. vRad reads over 67,000 stroke studies and 21,000 perfusion studies annually and has used that experience to train its NLP to "listen" for critical finding statements in real-time as its radiologists dictate their notes and their system can automatically call the referring physician with results to reduce time to treatment of these patients.

(continued)

---

[12] https://www.vrad.com/

[13] http://stroke.ahajournals.org/content/46/10/3020.long

[14] https://www.ncbi.nlm.nih.gov/pmc/articles/PMC4530420/

[15] https://www.ncbi.nlm.nih.gov/pmc/articles/PMC3088377/

To explore the potential for artificial intelligence in stroke imaging vRad partners with artificial intelligence vendors to explore the applications of deep learning in image recognition. vRad provided anonymized data from its clinical data warehouse to train the deep learning technology to search and "red flag" images that potentially show intracranial hemorrhaging (IH). vRad's PACS system can analyze incoming images such as the one shown in Fig. 13.7 and, using this technology, recognize potentially critical findings and triage the studies to be interpreted next by changing the normal order of the worklist as a result of a study whose critical nature demands immediate attention. The company also sees a potential future application of image based analytics/ deep learning to automatically populate the radiology report with specific findings that might be measurements, identification of specific pathology or even broader clinical impressions.

Rad is exploring the potential of using FHIR to achieve deeper EHR integration at both the health system and EHR vendor levels. One use case is to enable their PACS system to access clinical history (such as lab values) and outcomes (what actually occurred to a patient when a radiologist identifies a particular clinical finding). It also sees a role for FHIR in Query/Retrieve (Q/R). The DICOM standard for images we briefly mentioned earlier provides a trigger for automatic retrieval of prior image sets based on the arrival of a new image or the arrival of an order via an HL7 message. Unfortunately, the company says that it is not common that prior reports are accessible from the remote PACS by DICOM Q/R and HL7 messaging does not reliably support a Q/R interface for prior reports. Thus it sees FHIR as a way to reliably get access to a more complete clinical history on a patient, including prior imaging reports.

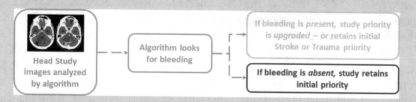

Fig. 13.7 If vRad's deep learning algorithm detects a hemorrhagic stroke in the image on the left their PACS platform can reprioritize the radiologist's worklist to upgrade the priority of that study reducing the response time back to the referring physician. (Courtesy vRad Radiology)

13.12 ResearchKit and ResearchStack

Apple's ResearchKit[16] and the Android ResearchStack[17] (developed with funding from the Robert Wood Johnson Foundation and maintained by developers from several companies and organizations) are open source frameworks that researchers can use to create a visual patient-facing module that explains the purpose of their study and obtains consent for participation, create surveys, and define visually attractive 'active tasks' in which study participants perform activities under semi-controlled conditions. Through integration with Apple's HealthKit, researchers using ResearchKit can access data such as daily step counts, calorie use, or heart rate recorded or measured by study participants on their phones. The ResearchKit and ResearchStack software are both freely available on GitHub.[18][19]

The Cancer Distress Coach (CaDC) mobile app was developed by the Duke Institute for Health Innovation and is an interesting example of the use of ResearchKit and ResearchStack. Work that led to this project was started years earlier by Dr. Sophia Smith who was trained in systems analysis and software development at IBM. She changed careers as a result of experiencing a second cancer diagnosis and earned a Masters of Social Work and a PhD and has subsequently done research on posttraumatic stress disorder (PTSD) in cancer survivors at the Duke Cancer Institute.

She found a PTSD mobile app in Apple's iTunes store and contacted its developers at the National Center for PTSD that is operated by the US Department of Veterans Affairs. They collaborated on spinning out the coaching component of that app out as a custom iOS app called Cancer Distress Coach (CaDC) and targeted cancer survivors. It was tested at Duke primarily with breast and lymphoma patients. A 2016 paper in Psycho-Oncology reported that "a large majority (89.7%) of participants endorsed at least moderate satisfaction with CaDC and 86.2% that it provided practical solutions to experienced problems. Also, 79.3% endorsed that CaDC was helpful in learning about and enhancing knowledge of posttraumatic stress. Most participants (72.4%) reported that CaDC helped to overcome the stigma of seeking mental health services." Importantly, about half of the study participants reported clinically significant reductions in PTSD symptoms after 4 weeks of app usage.[20]

Based on that success, the group received funding from Duke to move CaDC to ResearchKit and ResearchStack. These new versions of CaDC are available in both the Android Google Play and Apple iTunes app stores. In both versions the team reports that the standard framework facilitates the development of software that has the same look and feel of other research apps patients might be using through templates for functions such as obtaining patient permissions for research use of their data.

---

[16] http://researchkit.org/

[17] http://researchstack.org/

[18] https://github.com/researchkit/researchkit

[19] https://github.com/ResearchStack/ResearchStack

[20] https://onlinelibrary.wiley.com/doi/abs/10.1002/pon.4363

Fig. 13.8 CaDC's three modules are Learn which provides patient education; Insights that calculate each patient's personal PTSD risk score; and Activities that help users manage PTSD symptoms. (Courtesy Dr. Sophia Smith)

The CaDC app's three modules are illustrated in Fig. 13.8. They consist of three modules: **Learn** provides patient education about cancer related PTSD and what the clinicians who may be involved in a patient's care will do; **Insights** asks a series of questions to calculate each patient's personal PTSD risk score; and **Activities** that help users manage PTSD symptoms and include techniques such a guided imagery (therapist taking the patient through a scenario such as imagining you're walking through the forest, sit on a bench and invite anyone you want to have a conversation with you), soothing music and pictures, and relaxation and breathing exercises.

Results of the use of the ResearchKit/ResearchStack versions of CaDC were not published as of this writing but the team says that over 400 control/treatment patients are achieving results similar to those reported in the publication cited earlier and there are statistically significant differences between the control group that has access only to the Learn module and the treatment group that has access to all three modules.

13.13   Genomics Research Networks

As the cost of gene sequencing has dropped dramatically over the past few years there has been an explosion of interest in and potential for genomics research and the use of genomic data in clinical practice. Personalized or precision medicine is the individualized treatment of patients based on their own unique bio-genomic profile. While social, environmental and other factors clearly effect the development and course of disease, genomic variations are a major determinant of each person's health and response to treatments.

There are at least two major challenges that informatics applied to large genomic datasets can help solve. Many genomic variations are rare so aggregating data from multiple centers is important. It is often necessary to correlate these variations with clinical data to understand their ramifications. It is worth repeating that last point. We do not yet understand the ramifications of most genomic variations. By linking clinical data, where the phenotypic expression of genomic variations is stored, with the genomic data the hope is that we can understand the clinical manifestations of genomic variations.

Given the need for 'big data' about both genomic variations and clinical phenotypes, research networks may play an important role in advancing knowledge. For example, Boston Children's Hospital, Cincinnati Children's Hospital Medical Center and the Children's Hospital of Philadelphia, three major pediatric health systems, recently formed the Genomics Research and Innovation Network (GRIN) because each alone may have only a handful of patients with a particular rare disease. To help better understand the clinical ramifications of gene variations, GRIN seeks to create a broad database of annotated genomic and clinical data. It will add RNA, proteomics and metabolomics data to an existing shared library of genomic sequencing data. The network also wants to connect the genomic data with imaging results and electronic health records. The initial focus is on early childhood obesity, growth disorders/short stature, and epilepsy.[21]

The epilepsy project explores epileptic encephalopathy, a process where epileptic activity impairs overall brain function, including cognitive function, language, and behavior. Data from Online Mendelian Inheritance in Man (OMIM)[22] shows that a number of genes may have an association with this condition. Once exome sequencing (exomes are the protein coding genes in DNA) is completed, the data is to be stored in a joint, shared cloud environment combined with extracted phenotype data from the electronic medical records (EMR) of the three centers in an effort to better elucidate and understand these relationships.

13.14 Recap

We have now completed our journey through health informatics. You should recall that we began with the IOM's goal of a Learning Healthcare System that could aggregate and analyze data derived from actual patient care to develop new

---

[21] https://www.grinnetwork.org/

[22] https://www.omim.org/

knowledge and feed that knowledge back to the care of new patients. In this chapter we examined in some detail a few of the current approaches to the derivation of new knowledge from healthcare data. We have also seen how models built from that exploration of data can be instantiated into FHIR apps and other tools for use as patient care is delivered. While adoption has yet to be proven, we seem closer than ever to the goal.

Postscript

Given the scope of health informatics it is difficult to adequately describe the impact that the emergence of FHIR and its many applications is having and will have in the future in a book that is also both readable and aimed at a general audience. Hopefully, you now recognize that FHIR will encompass everything from clinical practice to insurance payments to patient engagement and research. As was my approach in the two predecessor volumes, I have tried to provide a broad view of the background that led to this new technology, the technology itself and its many current and potential future applications. Along the way I have provided information on the many web sites and publications that readers may use to continue to follow the rapid evolution of this fascinating new technology.

Of course, since FHIR will still be in active development at the time this publication appears and exploration of its applications has really only begun, it is likely that much of what I have written will change or otherwise become dated over the next few years. Indeed, there were a number of occasions when I felt the need to add new discussions or update what I had already written as a result of learning about applications of FHIR that came along as I was writing the book. Apple's support for a FHIR based personal health record is but one, but a potentially very impactful, example.

There are also a number of places where FHIR adoption by industry is incomplete. Virtually all EHR vendor FHIR implementations are limited to reading data. Cerner has released some ability to write data back to the EHR and hopefully other vendors will follow and the number or resources where this is supported will grow. Also, many EHR vendors are supporting SMART apps but only for patient facing use. This is because of the emphasis Meaningful Use places on patient access to their data from any app of their choosing. Hopefully, over time, support for provider apps for important use cases like clinical decision support will grow.

Inertia is of course also an impediment. There are many ongoing efforts started not that long ago that rest on technologies such as C-CDA whose significance will likely decline in the face of the more agile and workable FHIR approach. Hospital EHR vendors may have already invested substantially in the older IHE XCA

© Springer International Publishing AG, part of Springer Nature 2018
M. L. Braunstein, *Health Informatics on FHIR: How HL7's New API is Transforming Healthcare*, https://doi.org/10.1007/978-3-319-93414-3

interoperability specification and may be reluctant to start over. There can even be disagreements about what FHIR is! For example, is it sufficient to package data into the resource format and access it using a proprietary API? What about using the FHIR API to access data packaged in a document format that doesn't contain FHIR resources? I am optimistic all of this will get worked out and, given its momentum and superior usability, the full FHIR specification will prevail.

Despite these reservations, limitations and impediments, I believe this is an appropriate time to publish this book because it is now clear that FHIR will be accepted by the broad health informatics community and at least many of its key impacts and applications can be seen with reasonable accuracy from today's perspective. Clearly novel and unanticipated applications will continue to appear over the next few years and hopefully this book will provide you with the perspective to appreciate them.

A big part of my motivation was also to expand the universe of health care providers and information technology professionals who have access to a single volume through which they can become aware of FHIR and many of its key applications. I certainly hope that this book will be useful to those readers despite my obvious inability to accurately forecast the future.

Even as I was writing this book numerous changes have taken place in the rapidly accelerating FHIR landscape. As a result, I am even more certain than would be usual that it will contain errors and omissions despite my best efforts to avoid both. I have acknowledged the many topic experts who have helped by reviewing sections of the book but any remaining errors or omissions are, of course, entirely my responsibility.

Mark Braunstein
School of Interactive Computing
Georgia Institute of Technology
Atlanta, GA, USA
April, 2018

Useful Web Tools and Resources

For FHIR Developers	
HL7 FHIR Current Version	https://www.hl7.org/fhir/
Published Versions of FHIR	http://hl7.org/fhir/directory.html
FHIR Wiki	http://wiki.hl7.org/index.php?title=FHIR
FHIR Implementation Guides	http://www.fhir.org/guides/registry
Publicly Available FHIR Servers	http:// wiki.hl7.org/index.php?title=Publicly_Available_FHIR_Servers_for_testing
Open Source FHIR Implementations	http://wiki.hl7.org/index.php?title=Open_Source_FHIR_implementations
HAPI FHIR homepage	http://hapifhir.io/
clinfhir Launch Page	http://clinfhir.com/
SMART FRED FHIR Resource Builder	http://docs.smarthealthit.org/fred/
SMART Developer Documentation and Tools	https://dev.smarthealthit.org/
HL7 Da Vinci homepage	http://www.hl7.org/about/davinci/index.cfm?ref=common
The Argonaut Project	http://argonautwiki.hl7.org/index.php?title=Main_Page

(continued)

© Springer International Publishing AG, part of Springer Nature 2018
M. L. Braunstein, *Health Informatics on FHIR: How HL7's New API is Transforming Healthcare*, https://doi.org/10.1007/978-3-319-93414-3

FHIR Bulk Data Server Demonstration Tool	https://bulk-data.smarthealthit.org/
Follow FHIR	
FHIR Wiki	http://wiki.hl7.org/index.php?title=FHIR
Grahame Grieve's Blog	http://www.healthintersections.com.au/?page_id=208
David Hay's Blog	https://fhirblog.com/index/
HL7 FHIR Blog Index	http://wiki.hl7.org/index.php?title=FHIR_Blogs
FHIR Vimeo Channel	https://vimeo.com/channels/hl7fhir
FHIR News on Twitter	https://twitter.com/FHIRnews
John Halamka's Blog (not FHIR specific)	http://geekdoctor.blogspot.com/
Meetings	
HL7 FHIR DevDays	https://www.fhirdevdays.com/
HL7 FHIR Applications Roundtable	http://www.hl7.org/events/fhirapps.cfm
HL7 Partners in Interoperability	http://www.hl7.org/events/partners.cfm?ref=nav
HIMSS	http://www.himssconference.org/
App Galleries	
SMART Apps	https://apps.smarthealthit.org/
Cerner FHIR App Gallery	https://code.cerner.com/apps
Epic Orchard FHIR App Gallery	https://apporchard.epic.com/Gallery
Allscripts App Store	https://allscriptsstore.cloud.prod.iapps.com/
athenahealth Marketplace	https://marketplace.athenahealth.com/

Glossary of Terms and Acronyms

Not all of these terms are used in this book but you may encounter them as you further explore health informatics.

Accountable Care Organization (ACO) Medicare's outcomes-based contracting approach.

American Recovery and Reconstruction Act (ARRA) The Obama administration's 2009 economic stimulus bill.

Application Programming Interface (API) Specified formats for sending requests for information to a server and receiving the requested data back. See also REpresentational State Transfer (REST), a specific type of API used by FHIR.

Arden Syntax An approach to specifying medical knowledge and clinical decision support rules in a form that is independent of any electronic health record (EHR) and thus sharable across hospitals.

Area under the Curve (AUC) A measure of how well a model will rank a randomly chosen positive instance higher than a randomly chosen negative example.

Blue Button An ASCII text-based standard for heath information sharing first introduced by the Veteran's Administration to facilitate access to records stored in VistA by their patients. The newer Blue Button+ format provides both human and machine-readable formats.

Blue Button 2.0 A FHIR app platform developed by the Medicare program so that its beneficiaries can use their individual claims data with any FHIR app of their choice. It was released at HIMSS 2018.

Bulk Data Query Protocol (FHIR) A proposed new part of the FHIR specification allowing for the retrieval of specified data for a potentially large specified group of patients.

CDS Hooks An API for automatically invoking decision support from within an EHR workflow based on EHR embedded clinical logic.

Centers for Disease Control and Prevention (CDC) The federal agency focused on disease in the community.

© Springer International Publishing AG, part of Springer Nature 2018
M. L. Braunstein, *Health Informatics on FHIR: How HL7's New API is Transforming Healthcare*, https://doi.org/10.1007/978-3-319-93414-3

Centers for Medicare and Medicaid Services (CMS) The component of the Department of Health and Human Services that administers the Medicare and Medicaid programs.

Certificate Authority (CA) An entity that digitally signs certificate requests and issues X.509 digital certificates that link a public key to attributes of its owner.

Clinical Context Object Workshop (CCOW) An HL7 standard for synchronizing and coordinating applications to automatically follow the patient; user (and other) contexts allow the clinical user's experience to resemble interacting with a single system when the user is using multiple, independent applications from many different systems.

Clinical Data Repository A system that consolidates data from various clinical sources, such as an EMR or a lab system, to provide a full picture of the care a patient has received and as a platform for analytics and research.

Clinical Document Architecture (CDA) An XML-based markup standard intended to specify the encoding, structure and semantics of clinical documents.

Clinical Information Modeling Initiative (CIMI) An independent collaboration of major health providers to improve the interoperability of healthcare information systems through shared and implementable clinical information models.

CommonWell Alliance A group of major HIT companies that is working to achieve interoperability among their respective software products and services.

Complete EHR An EHR software product that, by itself, is capable of meeting the requirements of certification and Meaningful Use.

CONNECT ONC supported open source software for managing the centralized model of HIE.

Consolidated Clinical Document Architecture (C-CDA) The second revision of HL7's CDA architecture that attempts to introduce more standard templates to facilitate information sharing (a mandate of Meaningful Use Stage 2).

Continua A specification for interoperability of health and fitness devices. It is published by the Personal Connected Health Alliance.

Continuity of Care Document (CCD) An XML-based patient summary based on the CDA architecture.

Continuity of Care Record (CCR) An XML-based patient summary format that preceded CDA.

Cross-Enterprise Document Sharing (XDS) The use of federated document repositories and a document registry to create a longitudinal record of information about a patient.

Current Procedural Terminology (CPT) The American Medical Association's standard for coding medical procedures.

De-identified Patient Health Information PHI from which all data elements that could allow the data to be traced back to the patient have been removed.

Digital Imaging and Communications in Medicine (DICOM) A widely used standard for creation and exchange of medical images.

Direct A set of ONC-supported standards for secure exchange of health information using e-mail.

Domain Name System (DNS)   The naming system for computers, services or any resource connected to the Internet (or a private network). Among other things, it translates domain names (for example, eBay.com) to the numerical IP addresses needed to locate Internet connected resources.

EDI/X12   A format for electronic messaging that utilizes cryptic but compact notation primarily to support computer-to-computer commercial information exchange.

eHealth Exchange   A set of standards, services and policies that enable secure nationwide, Internet-based HIE using CONNECT or one of the commercial HIE products that support eHealth Exchange.

Electronic Health Record (EHR)   A stakeholder-wide electronic record of a patient's complete health situation.

Electronic Health Record Certification   A set of technical requirements developed by ONC that, if met, qualify an EHR to be used by an eligible professional to achieve Meaningful Use.

Electronic Healthcare Network Accreditation Commission (EHNAC)   An independent, federally recognized, standards development organization focused on improving the quality of healthcare transactions, operational efficiency and data security.

Electronic Medical Record (EMR)   An electronic record used by a licensed professional care provider.

Eligible Professionals (Medicaid)   Health providers who are eligible for Medicaid Meaningful Use payments: doctors of medicine, osteopathy, dental surgery, dental medicine, nurse practitioners, certified nurses, nurse-midwives and physician assistants who work in a federally qualified health center or rural health clinic that is led by a physician assistant.

Eligible Professionals (Medicare)   Health providers who are eligible for Medicare Meaningful Use payments: doctors of medicine, osteopathy, dental surgery, dental medicine, podiatry, optometry and chiropractic.

EMPI   An enterprise master patient index.

Extensible Markup Language (XML)   A widely used standard for machine- and human-readable electronic documents and the language used to define CDA templates. It is one of the representations of FHIR resources recognized in the specification.

F1 Score   A measure of the performance of a binary classifier using the weighted average of precision and recall.

Fast Health Interoperable Resources (FHIR®)   An HL7 initiative that seeks to use modern web standards and technologies to simplify and expedite real-world interoperability solutions.

File Transfer Protocol (FTP)   A standard network protocol used for the transfer of computer files between a client and server on a computer network. There is a sFTP version that provides security using encryption for files containing sensitive information.

GitHub   A hosting service for version control, mostly of software, using git, a system for tracking changes in computer files and coordinating work on those files among multiple people.

Health Information Exchange (HIE)   The sharing of digital health information by the various stakeholders involved, including the patient.

Health Information Service Provider (HISP)   A component of Direct that provides a provider directory, secure e-mail addresses and public-key infrastructure (PKI).

Health Information Technology (HIT)   The set of tools needed to facilitate electronic documentation and management of healthcare delivery.

Health Insurance Portability and Accountability Act of 1996 (HIPAA)   Legislation intended to secure health insurance for employees changing jobs and simplify administration with electronic transactions. It also defines the rules concerning patient privacy and security for PHI.

Health Level Seven (HL7)   A not-for-profit global organization to establish standards for interoperability.

Health Maintenance Organization (HMO)   An organization that provides managed healthcare on a prepaid basis. Employers with 25 or more employees must offer federally certified HMO options if they offer traditional healthcare options.

Health System   A network of providers that are affiliated for the more integrated delivery of care.

Healthcare Information Technology Standards Panel (HITSP)   A public and private partnership to promote interoperability through standards.

Healtheway   An ONC-supported public-private partnership to promote nationwide HIE via the eHealth Exchange. It was renamed The Sequoia Project in 2015.

HIMSS   Describes itself as "a global, cause-based, not-for-profit organization focused on better health through information technology (IT)."

HL7 Development Framework (HDF)   The framework used by HL7 to produce specifications for data, messaging process and other standards.

hQuery   An ONC-funded, open source effort to develop a generalized set of distributed queries across diverse EHRs for such purposes as clinical research.

Hypertext Transfer Protocol (HTTP)   A query-response protocol used to transfer information between web browsers and connected servers. HTTPS is the secure version.

i2b2 (Informatics for Integrating Biology and the Bedside)   A scalable query framework for exploration of clinical and genomic data for research to design targeted therapies for individual patients with diseases having genetic origins.

IHE Cross-Enterprise Document Media Interchange (XDM)   A standard mechanism for including both documents and metadata in zip format using agreed upon conventions for directory structure and location of files.

IHE Cross-Enterprise Document Reliable Interchange (XDR)   A standard mechanism for exchanging both documents and metadata using SOAP web services as the transport mechanism.

International Classification of Diseases (ICD)   The World Health Organization's almost universally used standard codes for diagnoses. The current version is ICD-10, but ICD-11 is scheduled for release in 2018.

International Health Terminology Standard Development Organisation (IHTSDO)   The former name of the organization that maintains SNOMED. It's now called SNOMED International.

Internet of Things (IoT)   In healthcare this describes a profusion of Internet-connected devices, sensors and other equipment that has the potential to transform care delivery.

Interoperability   The ability of diverse information systems to seamlessly share data and coordinate on tasks involving multiple systems.

IP Address   A 32-bit (the standard is changing to 128-bit to accommodate Internet growth) number assigned to each device in an Internet Protocol network that indicates where it is in that network.

JASON   An independent group of some 30–60 scientists that advises the United States government on matters of science and technology.

JavaScript Object Notation (JSON)   Is a relatively simple, human readable data-interchange format for packaging a group of data items that is also easy for computers to parse and generate. It is based on a subset of the JavaScript programming language that is widely used on the web.

Javascript Object Notation (JSON)   An open-standard file format that uses human-readable text to package and share data. JSON is one of (and the most commonly used) the data representations recognized in the FHIR specification.

JSON Web Tokens (JWT)   An open standard for securely sharing information as a JSON object.

Lightweight Directory Access Protocol (LDAP)   A protocol for accessing (including searching) and maintaining distributed directory information services (such as an e-mail directory) over an IP network.

Logical Observation Identifiers Names and Codes (LOINC)   The Regenstrief Institute's standard for laboratory and clinical observations.

Massachusetts General Utility Multi-Programming System (MUMPS)   An integrated programming language and file management system designed in the late 1960s for medical data processing that is the basis for some of the most widely installed enterprise health information systems.

Master Patient Index (MPI)   Software to provide correct matching of patients across multiple software systems, typically within a health enterprise.

Meaningful Use   A set of usage requirements defined in three stages by ONC under which eligible professionals are paid for adopting a certified EHR.

MEDCIN   A proprietary vocabulary of point-of-care terminology, intended for use in electronic health record systems (as a potential alternative to SNOMED-CT) maintained by Medicomp Systems.

Medicaid   The joint federal and state program to provide healthcare services to poor and some disabled U.S. citizens.

Medical Dictionary for Regulatory Activities (MedDRA) The International Conference on Harmonisation's classification of adverse event information associated with the use of biopharmaceuticals and other medical products.

Medical Logic Module (MLM) The basic unit in the Arden Syntax that contains sufficient medical knowledge and rules to make one clinical decision.

Medicare The federally operated program to provide healthcare services to U.S. citizens over the age of 65.

Modular EHR A software component that delivers at least one of the key services required of a Certified EHR.

Multipurpose Internet Mail Extensions (MIME) The Internet standard for the format of e-mail attachments used in Direct. S/MIME is the secure version.

National Drug Codes (NDC) The Food and Drug Administration's numbering system for all medications commercially available in the U.S.

ndjson A format for storing or streaming structured data that may be processed one record at a time. It is proposed for storing results for later processing in the FHIR Bulk Data Query Protocol.

OAuth 2 An open standard for access control, commonly used as a way for Internet users to grant websites or applications access to their information on other websites without giving them their passwords. It is a component of the SMART on FHIR specification.

Office of the National Coordinator for Health Information Technology (ONC) The agency created in 2004 within the Department of Health and Human Services to promote the deployment of HIT in the U. S.

Open mHealth A non-profit collaboration seeking to address interoperability in the mobile health app and device space.

OpenID Connect A simple identity layer on top of the OAuth 2.0 protocol, that verifies the identity of an end-user based on the authentication performed by an authorization server, as well as to obtain basic profile information about the end-user. OpenID Connect is a component of the SMART on FHIR specification. SAML is a competing technology for cross system user identity.

OpenNotes Is a national initiative (not a technology) working to give patients access to the visit notes written by their healthcare providers.

Outcomes-Based Contract An approach to pay for healthcare that rewards physician performance against certain defined quality metrics when combined with a lower-than-predicted cost of care.

Patient Portal A means (usually a web site or mobile app) for patients to access all of their health data stored in the electronic records of a health system.

Patient-Centered Medical Home (PCMH) A team-based healthcare delivery model often particularly focused on the management of chronic disease.

Pay-for-Performance (P4P) An approach to pay for healthcare that rewards physician performance against certain defined quality metrics.

Personal Connected Health Alliance (PCHAlliance) A non-profit organization formed by the Healthcare Information and Management Systems Society (HIMSS), Continua, a group of some 240 healthcare providers, communications, medical, and fitness device companies, and the mHealth Summit. It now publishes and promotes the adoption of Continua's interoperability framework.

Personal Health Record (PHR)   A web page app or other means through which patients maintain a consolidated record of their health data and other information related to their care.

Physician Group Practice (PGP) Demonstration   The first pay-for-performance initiative for physicians under the Medicare program.

Picture Archive and Communication System (PACS)   An archive of digital medical images (e.g. X-ray, CT, MRI) tagged with patient information with robust functionality built using the DICOM standard and support of viewing of images from remote locations.

Precision   The ratio of correct positive predictions to total positive predictions made by a binary classifier model. It is calculated as TP/(TP+FP) where TP is true positives and FP is false positives.

Preferred Provider Organization (PPO)   A network of providers who have contracted to provide care to patients (usually at a discounted price) under an insurance plan.

Primary Care Physician (PCP)   The generalist in a patient's care team who assumes overall responsibility for all their health issues and often the gatekeeper who must generate referrals to specialists.

Private Key   The protected (known only to its owner) part of the special pair of numbers used to encrypt documents using PKI.

Protected Health Information (PHI)   Health or health-related information that can be linked to or used to identify a specific patient. PHI is subject to strict HIPAA regulations.

Provider   Health professionals, including physicians, nurse practitioners, physicians' assistants, that are engaged in direct patient care.

Public Key   The public part of the special pair of numbers used to encrypt documents using PKI.

Public Key Infrastructure (PKI)   A widely used system for protection of documents, messages and other data that rests on a pair of public and private keys to allow for a variety of use cases.

Read Codes   A hierarchical clinical terminology system used in general practice in the United Kingdom being phased out in favor of SNOMED-CT.

Recall   A measure of the performance of a binary classifier based on the ratio of correct positive predictions to the entire class of those who had the result being predicted. It is calculated as TP/(TP+FN) where TP is true positives, and FN is false negatives.

Receiver Operating Characteristic (ROC) Curve   A graph that illustrates the diagnostic ability of a binary classifier system as its discrimination threshold is varied. The area under the curve (AUC) is equal to the probability that a classifier will rank a randomly chosen positive instance higher than a randomly chosen negative one.

Record Locator Service   Essentially a digital index (usually maintained by an HIE or HIN) that identifies where records are located based upon criteria such as a Person ID and/or record data type, as well as providing functionality for the ongoing maintenance of this location information.

Reference Information Model (RIM)   A pictorial representation of the HL7 clinical data (domains) that illustrates the life cycle of an HL7 message or groups of related messages.

Registration Authority (RA)   An entity that collects information for the purpose of verifying the identity of an individual or organization and produces a certificate request.

Representational State Transfer (REST)   Web interoperability principles proposed by Roy Fielding as a simple, consistent implementation of HTTP(S) basic commands (GET, PUT, POST or DELETE) for transfer of media (which can be data, images or other forms of digital information) between a server and a client. The ease and speed of REST development and led to its growing use for web interoperability. REST is FHIR's preferred transport protocol implementation for exchanging FHIR Resources.

Resource Description Framework (RDF)   A method for describing or modeling information on the web using subject-predicate-object expressions (triples) in the form of subject-predicate-object expressions that could be used to represent health ontologies (SNOMED, ICD-10). See Turtle as a means of representing RDF.

Security Assertion Markup Language (SAML)   Is an XML based single sign-on (SSO) login standard to authorized users for applications based on their sessions in another context.

Semantic Interoperability   The ability to share data that can be well enough understood by the receiving system that it can do things with that data that it might normally be able to do only with data it collected.

Semantic Web   The proposed next generation of web in which technologies like RDF would create a "web of data" in which browsers (and other tools) could "understand" the content of webpages.

Sequoia Project   An ONC-supported public-private partnership to promote nationwide HIE via the eHealth Exchange. Its name was Healtheway prior to 2015.

Service Oriented Architecture (SOA)   A software architecture in which services are provided to the other components by application components, through a communication protocol over a network. This was a predecessor to today's much simpler web services architecture using REST APIs.

Simple Object Access Protocol (SOAP)   A simple protocol for exchanging XML formatted information between applications using the Internet.

Simplified Mail Transport Protocol (SMTP)   Internet standard for e-mail used by Direct. The secure version is S/SMTP.

SMART on FHIR   A set of open specifications to integrate apps with electronic health records, portals, health information exchanges, and other health IT systems.

Synthetic Health Data   Facsimile clinical data created by a software system to realistically resemble actual patient data.

Systemized Nomenclature of Medicine (SNOMED CT) A comprehensive, hierarchical healthcare terminology system. It is now referred to as SNOMED CT and it is maintained by SNOMED International.

Templates The reusable basic XML-based building blocks of a CDA document that can represent the entire document, its sections or the data entries within a section.

Transition of Care Initiative (ToC) The effort to develop a standard electronic clinical summary for transitions of care from one venue to another.

Treatment, Payment or Operations (TPO) HIPAA exception for providers, insurance companies and other healthcare entities to exchange information necessary for treatment, payment or operations of healthcare businesses.

Turtle (Terse RDF Triple Language) A format for expressing data in the Resource Description Framework (RDF) data model by grouping three URIs to make a triple representing relationships such as a name is an author of a particular book.

Unified Medical Language System (UMLS) A service of the National Library of Medicine, it links many health and biomedical vocabularies and standards to facilitate interoperability.

Use Case A software and system engineering term that calls for developing a written description of how a user uses a system to accomplish a particular goal.

Veterans Health Information Systems and Technology Architecture (VistA) The VA's system-wide, MUMPS-based health information infrastructure.

View, Download, Transmit (VDT) A requirement of Meaningful Use Stage 2 that patients view, download or transmit their health information.

Web services A method of communicating between two devices or software applications over the Internet.

World Health Organization A United Nations concerned with international public health that also maintains the International Classification of Diseases (ICD) coding system for medical diagnoses.

X.509 Digital Certificate The technical name for an electronic document issued by a CA that uses a digital signature to bind a public key with an identity based on information from an RA.

XML See extensible markup language.

XMPI A cross-organizational master patient index capable of dealing with many unaffiliated hospitals and health systems.

Index

Printed in the United States
By Bookmasters